T0318718

COMPETENCY-BASED EDUCATION IN AVIATION

Kearns:
I dedicate this work to my parents, Eric and Dale Robinson,
who taught me that the price of success is hard work.

Mavin:
To my wife Karen and children Joshua, Alicia, Hugh and Amy.
Thank you again for your patience!

Hodge:
I dedicate this work to my wife Tammi and to
my children Sarahanne, Lachlan and Ashleigh.
You each inspire me.

Competency-Based Education in Aviation
Exploring Alternate Training Pathways

SUZANNE K. KEARNS
The University of Western Ontario, Canada

TIMOTHY J. MAVIN
Griffith University, Australia

STEVEN HODGE
Griffith University, Australia

LONDON AND NEW YORK

First published 2016 by Ashgate Publishing

2 Park Square, Milton Park, Abingdon, Oxfordshire OX14 4RN
52 Vanderbilt Avenue, New York, NY 10017

Routledge is an imprint of the Taylor & Francis Group, an informa business

First issued in paperback 2020

British Library Cataloguing in Publication Data
A catalogue record for this book is available from the British Library

The Library of Congress has cataloged the printed edition as follows:
Kearns, Suzanne K., author.
  Competency-based education in aviation : exploring alternate training pathways / by Suzanne K. Kearns, Timothy J. Mavin and Steven Hodge.
    pages cm
  Includes bibliographical references and index.
  ISBN 978-1-4724-3856-0 (hardback)
1. Aeronautics--Study and teaching. 2. Flight training. 3. Air pilots--Training of.
4. Competency-based education. I. Mavin, Timothy J., author. II. Hodge, Steven, author. III. Title.
  TL560.K429 2015
  629.132'52071--dc23

2015028316

ISBN 978-1-4724-3856-0 (hbk)
ISBN 978-0-367-66999-7 (pbk)

Contents

PART I: COMPETENCY-BASED EDUCATION IN AVIATION

PART II: AVIATION PROFESSIONAL TRAINING

List of Figures

List of Tables

Abbreviations

AIP	Aeronautical Information Publication
ALPA	Air Line Pilots Association
AMA	American Management Association
AME	aircraft maintenance engineer
AMM	aircraft maintenance manual
AMMTE	aircraft mechanics, technicians, and engineers
AMSL	above mean sea level
AQP	advanced qualification program
AR	augmented reality
ASG	acceleration-sensing glove
ATC	air traffic control
ATC1	Interview with "Air Traffic Control Training Expert, Africa and Europe"
ATC2	Interview with "Trainer, South East Asia"
ATC3	Interview with "Air Traffic Control Training Expert, South East Asia"
ATC4	Interview with "Air Traffic Control Training Expert, South East Asia"
ATCOs	air traffic controllers
ATO	approved training organization
ATPL	airline transport pilot license
ATQP	alternative training and qualification program
ATSEP	air traffic safety electronics personnel
BEI	behavioral event interview
CASA	Civil Aviation Safety Authority
CBE	competency-based education
CBT	competency-based training (as used in this book) also acronym for computer-based training
CBTE	competence-based teacher education
CC1	Interview with "Cabin crew trainer, South East Asia and Middle East"
CC2	Interview with "Cabin Training Expert, North America"
CC3	Interview with "International Organization Representative, North America"

CPL	commercial pilot license
CRM	crew resource management
CTA	cognitive task analysis
DACUM	"Developing a curriculum"
DG	dangerous goods
DME	distance measuring equipment or designated medical examiner
EASA	European Aviation Safety Agency
EBT	evidence-based training
EDA	electrodermal activity
e-learning	electronic learning
ENG1	Interview with "Training manager, South East Asia"
ENG2	Interview with "Training manager, South East Asia"
ENG3	Interview with "Training manager, South East Asia and Europe"
ENG4	Interview with "Maintenance Training Expert, Europe"
EP	emergency procedure
FAA	Federal Aviation Administration
FCLTP	Flight Crew Licensing and Training Panel
FDR	flight data recorder
FFS	full flight simulator
FIR	flight information region
FMS	full mission simulator
FNPT	flight navigation and procedures trainer
FOR	frame of reference
FPS	flight progress strips
FSTD	flight simulation and training device
Ft	feet
FTD	flight training device
GPS	global positioning system
HF	human factors
HRD	human resource development
HRO	high-reliability organization
HRT	high-reliability theory
HTA	hierarchical task analysis
IATA	International Air Transport Association
IBSTPI	International Board of Standards for Training, Performance, and Instruction
ICAN	International Commission of Air Navigation
ICAO	International Civil Aviation Organization
IMC	instrument meteorological conditions
IR	instrument rating
ISD	instructional system design

ITF	International Transport Worker's Federation
ITQI	IATA Training and Qualification Initiative
ITS	intelligent tutoring system
KSA	knowledge, skill and attitude
LAME	licensed aircraft maintenance engineer
LMS	learning management system
LOFT	line oriented flight training
LOS	line operational simulation
MAPP	model for assessing pilots' performance
MEL	minimum equipment list
m-learning	mobile learning
MM	maintenance manual
MPL	multi-crew pilot license
NGAP	next generation of aviation professionals
NOTECHS	non-technical skills
NTS	national training system
NVQ	National Vocational Qualifications
OJT	on-the-job training
PANS	procedures for air navigation services
PBE	protective breathing equipment
PBTE	performance-based teacher education
PIC	pilot in command
PIL1	Interview with "Training manager, South East Asia"
PIL2	Interview with "ICAO (retired), Europe, North America and South East Asia"
PIL3	Interview with "ICAO (retired), North America"
PIL5	Interview with "Training manager, South East Asia"
PIL6	Interview with "Training manager, South East Asia"
PIL7	Interview with "Training manager, South East Asia"
PIL8	Interview with "Airline Training Expert, Asia"
PIL9	Interview with "Instructional Designer, North America"
PPL	private pilot license
PTT	part task trainer
ROI	return on investment
SA	situation awareness
SARP	standards and recommended practice
SOE	schedule of experience
SPM	standard practices manuals
TAFE	technical and further education
TEM	threat and error management
TPO	terminal proficiency objective

VFR	visual flight rules
VMC	visual meteorological condition
VR	virtual reality
ZFT	zero flight time

Acknowledgments

This work would not have been possible without the involvement of aviation experts who generously donated their time for interviews and pre-publication reviews of this manuscript. The authors of this text would like to express sincere appreciation to these professionals.

PART I
Competency-Based Education
in Aviation

Introduction to Part I
Competency-Based Education in Aviation

> I don't think anybody was excited about competency-based training, I really don't.
> I think it was this almost gradual evolution. I don't recall any point at which there
> was this conscious decision, "okay we're starting to switch to competency-based
> approaches!" ... It was just something that was gradually happening in the various
> different countries. (ATC1)[1]

The term *competency-based education* (CBE) has become increasingly common
within aviation in recent years. In the authors' experience, almost everyone involved
in training and assessment has their own unique opinion. The views expressed
range from "it just does not encapsulate the complexity of performance" to "well,
it appears to be a logical approach" and sometimes to an almost evangelistic
enthusiasm for CBE. What has led industry, professional and regulatory groups
to ask for competency-based education? Was their decision based on traditions
of educational instruction, or on a new understanding of how adults learn and
perform in the workplace?

Hours or Competence?

CBE initially caught on within aviation training because it was generally
recognized that a focus on "hours" of training (such as a pilot's flight hours)
did not necessarily reflect proficiency or skill. It seems logical that what occurs
during those hours is more important than the hours themselves. Linking length of
training (hours) with the quality of a person's skill is a bit like saying the length of
a book is directly related to its quality—in which case the authors of this text may
consider adding a few additional chapters.

Why, then, have we historically equated *hours* of training with *competence*?
The aviation industry has a long history of regulating a specific number of training
hours per license or certificate. Similarly, as challenges have arisen in our industry
and new training needs are identified, we have addressed these issues by "throwing
a few hours of training" at the problem (PIL3).[2]

However, thanks to new training technology and a better understanding of
instructional methods, the industry has come to recognize that we can improve
upon the hours-based methods of our professional predecessors. CBE has gained
popularity as the approach that may lead to training improvements.

---

1 ATC1: Air Traffic Control Trainer, Southeast Asia.
2 International Regulator, retired, North America.

What is Competency-Based Education?

This is a question without a simple answer. Most aviation professionals vaguely understand that CBE focuses on the *quality* of training rather than *number of hours*. While writing this book, the authors conducted interviews with a variety of aviation training professionals from around the globe. Interestingly, each professional had a unique understanding of competency-based education. The differences among them tended to be related to their role in the industry:

- Instructional designers understand CBE as student-centered learning that is tailored specifically to the needs of each learner and their professional role.
- Professionals who work closely with multi-crew pilot licensing (MPL; further discussed in Chapters 2 and 4) understand CBE as the application of MPL training.
- Regulators regard CBE as an evolved form of training that incorporates modern teaching methods. Yet they acknowledge some challenges; for example, CBE can be expensive to implement, competency-based regulations must be adapted differently depending upon the organizational context, and many instructors might not fully understand how to modify teaching to be competency-based.
- Instructors with practical experience in teaching CBE often describe it as scenario-based training that they facilitate by coaching the learner.
- Recruitment and selection professionals regard "competencies" as a listing of attributes that can be used to select appropriate candidates for a job.

Ultimately, all of these experts are correct in different ways. To help clear up any confusion, we propose the following definitions, which will be explained and expanded upon throughout this book:

Competence: the ability to fully participate in a complex social practice, such as an aviation profession. Full participation requires skills, knowledge and attitudes relevant to that practice.

Competencies: negotiated and agreed written statements (texts) that attempt to represent the ability to fully participate in a social practice. Competencies embody assumptions about:
- the nature of competence
- the number of discrete written statements (texts) required to represent competence
- the most appropriate language for representing competence.

Competency-based education (CBE): instructional design, training, and assessment that systematically references written competencies.

To assist with understanding a large and complicated issue, we have compiled some key features of these three elements in Figure P1.1.

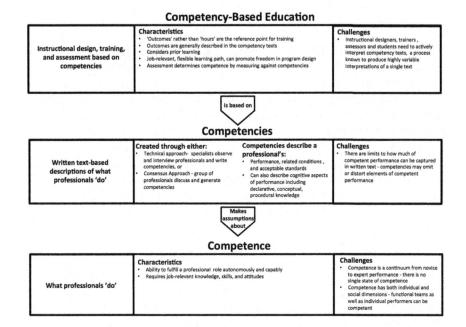

Figure P.1   Key Concepts of Competency-Based Education (CBE)

CBE Considerations

Despite the recent widespread popularity and adoption of CBE within the aviation industry, the approach is not free from criticisms. Like any methodology, it has strengths and weaknesses, which will be explored in detail throughout this book. The following concerns serve as an introduction to some of the issues associated with CBE.

Reductionism

If we take something as complex as expert professional practice and convert it into a textual statement, it can become artificially simplistic and lose aspects of real-world operations. The risk with this is that narrow competencies may dominate the curriculum. Such an approach can result in poor training by encouraging learners to demonstrate knowledge that fulfills a "checklist," instead of encouraging them to excel and to think critically (Leung 2002). A narrow competency model may limit the intuition, experience, reflection and higher-order competence that are necessary for expert, comprehensive or well-established practice (Talbot 2004). We must ensure that CBE considers non-technical professional attributes (problem-solving, teamwork and decision-making) and the interconnectedness of competencies with one another.

Interpretation

When text-based written competencies are produced, even if they are of excellent quality, their application will rely on individual interpretation. Throughout this book, the term hermeneutics is introduced to describe the skill of *text interpretation*, which is an important component of CBE. According to hermeneutic theory, different groups of people in the industry will interpret competency texts in different ways. Students who have a limited understanding may try to conform exactly to what the text says. Regulators may focus on how performance can be standardized within the competency framework and how to assess competencies that are more gray than black and white. Instructors who are responsible for converting competencies into training curricula may have varying levels of understanding of the same competencies.

Personalized Training

A focus on competencies, rather than on time served in training (hours), results in a system that allows learners to progress at their own pace. This individualized, flexible approach to training may lead to improved learner motivation, better skill development and reduced training duration (which results in cost savings). The length of training is still an important consideration of CBE, but training time is seen as a resource, rather than as an organizing framework for instruction.

End Goal

It is important to recognize that competence is a point on a the path towards expertise and does nto represent the end of a learner's development. Talbot (2004) presents an interesting framework for understanding how "competent" performance is related to the development of professional expertise:

- competence: a trainee "knows how"
- developing proficiency: a professional "shows how"
- global development: a professional "does"
- expertise: an expert "does well."

Therefore, competence is not the same as the development of expertise that would be acquired during the course of professional practice. CBE focuses its curriculum on what learners need to be able to do in order to perform capably and autonomously in real-world operations. Competence is not an indication of mastery. There is no avoiding the fact that expert mastery requires real-world experience and a great deal of time.

Standardization

CBE is based upon a standardized understanding of the knowledge, skills and attitudes required by a professional within a given field. The list of competencies is standardized and often it is generated by international regulators, which creates a uniform understanding around the globe. The advantage of this is that it should standardize training practices on a global scale. However, the challenge is that it may undervalue personal reflection, responsibility and ideals. Standardization can be too limiting—it can cause trainees to only be understood in terms of conformity and sameness (Leung 2002). It can also result in a focus on the minimum acceptable performance standards.

Administrative Burden

Regulatory oversight of traditional training programs, often referred to as a "check the box" approach, is accomplished by (1) organizations creating training materials that cover specified topic areas and (2) the regulator reviewing that curriculum. This can be considered a "black-and-white" approach, because the regulator checks to see whether *all* the content areas are included in the curriculum, and if some are missing, the airline or training company must revise the curriculum. By comparison, CBE is "gray": the training must result in learners achieving competency, but this can be accomplished through a wide variety of methods, equipment and instructional strategies. This can create a significant administrative burden and it requires a highly sophisticated regulator to implement and oversee the training.

Bias

In aviation, competencies are written through either a process of observing and documenting professional behavior on the job or through consensus of a group of experts who discuss and agree upon what defines competence. With either of these approaches, it is important to understand that the creation of competencies is not free from individual values; the meaning of the competencies is shaped by the people who create them (Leung 2002). Experts may not be conscious of the full range of their professional competence, as job elements may become intuitive and not consciously accessible or easily described. Professional groups, regulators, employers and learners may have different ideas about what aspects of the work are most important. This process is political to some extent, because it allows regulators to influence what will be included as important professional competencies (Leung 2002).

Competency-Based Education in Aviation: Book Overview

This book is organized into four parts. Part I, introduced in this chapter, contains Chapter 1, which reviews the history of CBE in the general educational system, and Chapter 2, which explores the origins of CBE in aviation. Part II presents insights into the history of training within aviation professions (air traffic control, pilots, cabin crew and maintenance engineers) as well as the present-day incorporation of CBE within these fields. Part III discusses CBE in relation to best practice, including how competencies are written, how to create training curricula based on competencies and how assessments are conducted. Part IV explores the role of teaching technologies and future considerations for CBE.

Expert Interviews

As CBE is a relatively new consideration within aviation training, the authors of this book turned to industry experts for insight into how competencies impacted training practices in the disciplines of air traffic control (ATC), piloting, cabin crew and maintenance engineering. Experts from various regions around the globe shared their experiences, opinions, praise and criticisms of CBE in interviews. Excerpts from these interviews are included throughout the book and are the foundation of the profession-focused Part II of the book (which details how training is conducted in each discipline and the role of CBE). For the purpose of anonymity, the experts interviewed are referred to by their professional role and geographic region only.

This work would not have been possible without the involvement of experts who donated their professional time for the interviews and for their pre-publication reviews of this manuscript. The authors of this text would like to express sincere appreciation to these professionals for contributing their time and expertise to this work.

Chapter 1
What is Competency-Based Education?

Questions Considered in this Chapter:

- What are the origins of competency-based education?
- What are the main traditions within the CBE movement?
- What are the essential features of CBE?
- What are strengths and challenges of the approach?

Competency-based education (CBE) (also referred to as "competency-based training") has been adopted by industries, professions and national training systems all over the world, and the aviation industry is no exception. The aviation industry has been using CBE to train skilled personnel for more than a decade. The implementation of CBE approaches has proceeded on the assumption that traditional forms of training have fixated on hours (a "time served" model) and that the training curriculum is not attuned to the real world. Changes within the aviation industry itself have also favored CBE. As the industry and its regulators embrace the values of transparency and accountability, training for industry roles is more subject to scrutiny than ever before. CBE appears to offer industry stakeholders a higher level of transparency than traditional approaches, reinforcing the case for training reform.

As a result, aviation industry newcomers experience training programs that are generally shorter and less expensive than their predecessors. These programs are based on *competencies*, which are specially formatted written statements/ texts designed to capture and encode the real, observable tasks (competence itself) that pilots, maintenance engineers, air traffic controllers and cabin crew undertake in their daily work. Competency documents are appraised and endorsed by industry stakeholders. Trainers, assessors and program designers are obliged to systematically reference their work to the competency documents. This system is designed to ensure that training and assessment is tightly aligned to the accepted realities of the industry. At the same time, a focus on pre-specified competencies allows training to pursue the most efficient means of skill development rather than being bound to complete specific numbers and types of training hours. This principle ensures, at least in theory, that faster learners are not slowed down and that slower learners are not completed prematurely. CBE thus offers the aviation industry accountable, relevant and efficient training in a new era focused on increasing value and lowering risk.

Where Competency-Based Education Comes From

The aviation industry is not alone in its adoption of CBE. In many jurisdictions, training for roles in engineering, psychology, education, medicine and law has become competency-based. The reasons include transparency in risky environments, cost control and consistent training output. The wide adoption of CBE raises the question of where the approach came from and exactly what it is. There is some debate about the origins of CBE, but it appears that two different traditions of CBE can be identified, which account for much of the diversity within the field today.

The term *competence* started to be used in academic debates in the 1960s. Bernstein (2000) explains that social scientists became interested in innate, generic abilities such as language use, practical reasoning and communication as a way to explain common human achievements. The concept of competence introduced a new way to talk about learning that stressed the significance of everyday experiences for triggering and shaping learning. In some contemporary settings, such as human resource development (Boyatzis 2007), CBE is clearly related to this notion of competence.

But in other contemporary industry and system settings, a peculiar shift in terminology is responsible for the prevalence of the term *competency*. For example, in aviation it means a specific ability to do something that can be defined in advance. Bernstein (2000) distinguished this kind of approach, which he described as a "performance" approach, from the idea of generic competence. In Bernstein's view, the performance mode is oriented to external tasks, focuses on observable behaviors and is generally favored in situations where education and training are deployed for some system or industry-specific ends.

The competence and the performance traditions of CBE will be described in the next two sections of this chapter. Note that although they are treated separately here, contemporary examples of CBE often blend these two traditions.

The Competence Tradition of CBE

The idea of competence made its way from the social sciences into education and training during the 1960s and it refers to "procedures for engaging with, and constructing, the world. Competences are intrinsically creative and tacitly acquired in informal interactions. They are practical accomplishments" (Bernstein 2000, 42). In addition, competence can be acquired by anybody, learners are "active and creative in the construction of a valid world of meanings and practice" (43), and competence is inherent in the working of the human mind. Bernstein mentions the ideas of linguistic competence (Chomsky 1965), cognitive competence (Piaget 1952), cultural competence (Lévi-Strauss 1963), member's competence (Garfinkel 1967) and communicative competence (Hymes 1966).

Piaget's ideas about learning influenced the work of psychologist Robert White, who made a seminal contribution to CBE. In his "Motivation Reconsidered: The Concept of Competence," White (1959) proposed that the need to be *competent* should be regarded as an important type of human motivation. He defined competence as "an organism's capacity to interact effectively with its environment." Another psychologist, David McClelland (1973), later took up White's lead and argued that the concept of competence should replace that of "intelligence" as a central focus in psychology. The challenge that occupied many psychologists at the time was how to test and predict suitability for particular jobs. McClelland proposed that tests be based on occupation-specific activities rather than abstract puzzles; be related to life outcomes such as communication skills, patience and goal-setting; and assess not only the ability to provide "correct" answers but also creative problem-solving abilities.

In the 1970s, the American Management Association (AMA) and McBer and Company (with input from David McClelland) undertook a large-scale study to determine the "competencies" of effective managers (Hayes 1979). This work was influential in the field of human resource development (HRD). Through the advocacy of Patricia McLagan (1980) and others, the competency model was used to conceptualize recruitment, training and development requirements. McLagan explained that "Competency models are decision tools which describe the key capabilities required to perform a job" (23) and that users of these models "must be able to recognize examples of competence and notice missed or mismanaged opportunities for competence. This may require training in observation and questioning skills" (26).

The emphasis here on underlying ability, and the research needed to recognize it, reinforces the link between the new HRD movement and the competence mode described by Bernstein (2000). It also illuminates the gap between underlying ability (McLagan 1980) and performance-oriented approaches to teacher education and vocational training that were adopted in the 1970s and 1980s.

Richard Boyatzis (1982), in *The Competent Manager*, built on the work of the AMA (Hayes 1979) and McLagan (1980):

> A job competency is an underlying characteristic of a person in that it may be a motive, trait, skill, aspect of one's self-image or social role, or a body of knowledge which he or she uses. The existence and possession of these characteristics may or may not be known to the person. … Because job competencies are underlying characteristics, they can be said to be generic. A generic characteristic may be apparent in many forms of behavior. (Boyatzis 1982, 21)

Boyatzis (1982) emphasized the need for sophisticated methodologies to determine competences for management development. His model describes effective job performance as "congruence" between individual competency, the job's demands and the organizational environment. Competency is a component in a complex system, and competent performance is dependent on features of the job and organizational context.

The competence tradition of CBE can also be seen in psychology (Kaslow et al. 2007), nursing (Scott Tilley 2008), and engineering (Patil and Codner 2007), and also some national training systems—for example, in France (Brockmann et al. 2008). These forms of CBE share an emphasis on the underlying, personal nature of competence and the one-to-many relationship between competence and particular competent performances.

The Performance Tradition of CBE

The performance tradition of CBE (using "performance" in Bernstein's sense) is common in education and training systems that are focused on initial preparation for skilled occupations. In the context of the Cold War, the launch of Sputnik in 1957 by the USSR triggered a crisis in American self-confidence. The education system came in for criticism, and performance-based teacher education (PBTE) was one of the measures adopted to restore American technological superiority. This new approach was based on the concept of *behavioral objectives*, which are written statements/texts that *describe* desired performance, indicate the *level of performance* and state the *conditions* under which the performance is to be demonstrated. Using the behavioral objectives format, desired teacher performances were analyzed, described and documented for use by program designers, teacher educators and examiners. PBTE was supported by government funding programs and was widely adopted in teacher education colleges throughout the United States (Hodge 2007).

PBTE soon became more than just the application of behavioral objectives theory to task analyses of desired performances. Stanley Elam (1971) described the essential characteristics of PBTE as follows:

1. "Competencies (knowledge, skills, behaviors) to be demonstrated by the student are
   - derived from explicit conceptions of teacher roles,
   - stated so as to make possible assessment of a student's behavior in relation to specific competencies, and
   - made public in advance;
2. "Criteria to be employed in assessing competencies are
   - based upon, and in harmony with, specified competencies,
   - explicit in stating expected levels of mastery under specified conditions, and
   - made public in advance;
3. "Assessment of the student's competency
   - uses his performance as the primary source of evidence,

- takes into account evidence of the student's knowledge relevant to planning for, analyzing, interpreting, or evaluating situations or behavior, and
- strives for objectivity;

4. "The student's rate of progress through the program is determined by demonstrated competency rather than by time or course completion;

5. "The instructional program is intended to facilitate the development and evaluation of the student's achievement of competencies specified." (Elam 1971, 6–7)

By the 1960s, theories of "mastery learning" (Bloom 1968) and "criterion-referenced" assessment (Glaser 1963) had been incorporated into the concept of PBTE. Mastery learning and criterion-referenced assessment were revolutionary developments in education and training. Mastery learning was originally proposed for use in apprenticeships (Kornhauser 1922) and then suggested for use in schools (Carroll 1963). It was refined and promoted by Bloom (1968) as an egalitarian education method that was appropriate to a democratic society. It commits to giving all learners a chance to achieve the goals of their education by providing whatever time and resources are necessary for all learners to succeed. The corollary of this approach is that the amount of time spent in a program is not a relevant measure of achievement. Behavioral objectives theory reinforces this point by emphasizing the outcome rather than the training or education process.

Criterion-referenced assessment can also be described as egalitarian (as opposed to norm-referenced assessment, which is based on comparing the results of different learners to determine achievement). Criterion-referenced assessment judges achievement by the extent to which a learner's performance resembles a previously determined criterial performance. It complements a mastery learning approach by allowing any learner who has mastered the objectives of learning to pass. According to Elam (1971), transparency and accountability are also essential characteristics of PBTE, meaning that the goals, processes and assessment of any PBTE program can be subjected to public scrutiny.

During the 1970s, the term *performance-based teacher education* gave way to *competence-based teacher education* (CBTE). In his description of the essential characteristics of PBTE, Elam (1971) was already using the language of competence, which he defined as knowledge, skills and behavior. This definition reflects the idea that learning acquired through training is internal to the learner, but can be observed in behavior or performance. However, according to Bernstein's (2000) distinction between performance and competence modes of education, Elam's model remains a performance approach, with the outcomes of training explicitly tied to observable tasks rather than an internal capability.

Essential Elements of CBE

Despite the differences in the way CBE is conceptualized in the competence and performance traditions, contemporary CBE approaches share some essential elements. These include assumptions about the nature of competence and its role in education and training; the use of written statements/texts to represent competence; and the importance of these competency texts in instructional design, training, assessment and credentialing. CBE approaches display variation in each of these elements, resulting in a wide range of CBE.

The Nature of Competence

There are several challenges associated with understanding competence, which is partly due to changes in the way we think about education and training. In contemporary education and training for occupations, the focus is moving away from traditional representations of subject matter and toward competent performance. For example, pilot training moves away from traditional classroom subjects to be memorized and then demonstrated on a multiple-choice test towards scenario-based instruction where learners must apply several categories of subject matter towards realistic contexts. The current chapter has discussed two ways of regarding competence and the traditions of education and training associated with them. Both of these approaches to CBE share the assumption that the content, delivery and assessment of training should be oriented to the goal of full, skillful participation in an occupational practice. *Practice* is understood as an organized endeavor with some history, which is maintained and developed by a community of practitioners. The mastery of a body of skills, knowledge and attitudes distinguishes the full participant in the practice. This conceptualization of competent performance within a practice is drawn from Lave and Wenger's (1991) theory of situated learning. For Lave and Wenger, learning is a function of progressive immersion in, and successful adaption to, a social practice or occupation.

Traditional conceptualizations of education for occupations focus on the development of knowledge that has been abstracted from practice settings, with the expectation that the learner will later apply the acquired knowledge to practice. In CBE, knowledge and application are taught together. Learners are exposed to knowledge, skills and attitudes in context, and are assessed on their ability to participate in a practice in a skilled way.

Competency Texts

An essential feature of CBE is the use of written statements/texts that represent competent participation in a practice. These texts may be called *competencies* or *standards*. Their production is generally an elaborate process, conducted by bodies that have the power to determine the legitimate representations of occupational practice (within aviation this takes place at the regulatory level). There are two basic models of identifying and agreeing on competencies: the technical model and the consensus model. The *technical model* uses analytic techniques to objectively identify the essential features of competencies. Although the origins of the technical model are not clear, the work of American business thinker Frederick Taylor is undoubtedly formative. Taylor (1904/1964) developed the concept of *scientific management*, in which managers would first gather knowledge of work processes from workers and then analyze the processes to determine more efficient ways to do the work. One of Taylor's key innovations was to think of processes in terms of *tasks*, rather than the day work and piecework schemas that had been used for calculating work up to that point. By analyzing work into discrete tasks, managers would be better placed to understand and redesign processes. Gilbreth and Gilbreth (discussed in Kirwan and Ainsworth 1992) developed the use of time and motion studies and process charts to enhance the managers' analyses.

The alternative model for identifying and agreeing on competencies is the *consensus model*, in which consensus (rather than analysis) is used to guide the identification of competencies. An example of the approach goes by the acronym DACUM, from "developing a curriculum," which grew out of a Canadian government initiative to address unemployment in the 1960s. This initiative required a process for rapidly analyzing jobs for vocational training programs, and a method for identifying competencies was developed in 1967. The DACUM method works on the premise that "people employed in various occupations are capable of describing the knowledge, skills and attitudes required to work in them" (Joyner 1995, 248). It involves a specially trained facilitator working with a group of about 6 to 12 workers and/or supervisors who are regarded as competent within their given occupation. The process traditionally takes a few days and starts with brainstorming the duties and tasks that comprise the work. The full traditional DACUM process involves subsequent phases in which specialists further analyze the tasks and develop instructional plans and materials to create a full curriculum package. A less formal implementation of the consensus model gets occupational experts together to debate and argue in order to determine competencies. This process may or may not be facilitated by a specialist in competency development. At the end of the process, the contributors more or less agree that the resulting competencies capture the essentials of the occupation in question.

The technical and consensus models of competency identification reflect different assumptions about occupational knowledge. The technical approach presents a more "scientific" stance, which supposes that experts can only describe their competence in a limited way, and that an alternative (or additional) lens is necessary to provide a valid representation of competence. The consensus approach is more pragmatic and views competence as essentially transparent. It assumes that workers and supervisors possess conscious knowledge of—and capacity to communicate—what makes for competent work. The validity of a consensus model in a given situation depends on the level of consensus that can be demonstrated. Chapter 7 explores the process of developing competency texts in more detail.

Whether a technical or consensus approach is used, the outcome is always written statements/texts. These texts are critical for CBE as a mechanism for sharing and controlling knowledge. The texts are the key means of communicating features of competent work that have been identified, agreed upon, described and analyzed. The form of the competency texts varies across contexts and is dependent on a range of factors. For example, some texts may follow the principles of behavioral objectives theory, which aims for aligning the written competencies with observable occupational activities. In contrast, some texts record less tangible information such as "knowledge" or "value" requirements for a role. Competency texts also vary in terms of how detailed they are. They may be highly structured and contain many components or they might highlight just a few broad categories of information.

One particular approach to structuring competency texts has been widely adopted in implementation of CBE. *Behavioral objectives theory* assumes that texts can be designed to transmit the writer's intentions clearly and precisely. Robert Mager (1962), who popularized the behavioral objectives technique, explained that desired performance must be described in terms of observable behavior. In addition, the level of performance must be specified and the context in which the performance takes place must be adequately described. An effective behavioral objective will therefore include a descriptor and will specify a criterion or criteria of performance and an indication of the conditions of the performance. Mager argued that objectives that adequately specify these types of information would transmit the writer's expectations to the reader implementing the competencies. Many competency texts are based on the behavioral objectives model (e.g., Australian Training Package competencies).

Competency statements/texts may include categories of information that are believed to enable competent performance, such as knowledge, skills, attitudes, values and understanding. In a strict behavioral approach to writing competencies, statements of knowledge and skills are not considered valid because possession of them cannot be directly observed. However, sometimes these concepts are combined with a behavioral approach, producing a hybrid format.

The process of developing competency texts also reflects assumptions about the level of segmentation appropriate for the purpose of representing competent work. CBE systems embody some rationale for dividing up competence to yield competency texts—a few broad competencies or a greater number of detailed competencies. Approaches that align with Bernstein's (2000) competence model may opt for fewer texts to represent competence, while those aligning with the performance model may produce more.

Education and Training for Competence

Written competency statements/texts are used for a range of purposes, including recruitment, work design, professional development, entry-level training, assessment and performance management (Gonczi, Hager and Oliver 1990). The main uses appear to be initial and continuing education, and training and assessment processes. In these cases, competency texts serve as a reference point for program design and development, learning resource development and the assessment design. Learning and assessment are derived from competency texts by designers, developers, trainers, teachers and assessors. Training to undertake this work varies across industries, ranging from specialized induction-style programs used in some settings to prepare people to use the texts, through to more informal arrangements involving newcomers working alongside experienced workers to develop the skills.

A key assumption of CBE is that the meaning of competency texts is transparent to users. Elam's (1971) description of CBE highlights the publicly accessible nature of the information in the competencies, and this message was also promoted by Mager (1962) and other proponents of behavioral objectives. The DACUM process also assumed that competent professionals are inherently able to communicate the knowledge, skills and attitudes that make up competent performance in their professional role. Research suggests that in at least some implementations of CBE, the process of translating competency texts into program content, delivery approaches and assessment instruments is a sophisticated skill that may take some time to develop. According to research by Hodge (2014), it took trainers and assessors about a year to develop the ability to interpret competency texts in the Australian competency-based vocational education system. Prior expertise in the skills addressed by the texts was not sufficient preparation to handle the challenge of decoding the information of competency texts.

CBE thus involves assumptions about competence and its relation to education; processes for deriving appropriate language to represent competent practice; and the use of the texts to guide instructional design, training and assessment. The ability to interpret the intention of competencies through the written language is crucial for trainers and assessors who implement CBE. In addition, implementing designers, trainers, assessors and/or bodies external to the education and training system may be involved in validation, moderation, auditing and evaluation processes. It should be noted that even though CBE is oriented to practice, the central role of written statements/texts separates CBE from situated learning processes. For Lave and Wenger (1991), the codification of practice knowledge—and the use of codified knowledge to determine educational processes—is itself a mark of traditional approaches to education.

Benefits and Challenges of CBE

Advocates of CBE argue that the approach has distinctive benefits. Some of these benefits are stated in terms of the approach itself and some contrast CBE with traditional or alternative approaches. One benefit consistently identified by advocates is that CBE focuses on outcomes rather than inputs (Harris et al. 1995). This benefit is sometimes explained through contrasting CBE with traditional "time served" approaches that focus on hours of training completed. In a CBE system, the education and training stop when the learning objectives have been achieved and learners are not detained for longer than necessary. Meanwhile, less able learners remain in the program until they have mastered the content. CBE thus offers a more efficient model of learning than approaches that commit learners to fixed training periods.

Another benefit is that CBE systems have the potential to take prior learning and previously developed competence into account (Smith and Keating 2003). If learners can demonstrate that they already possess outcomes sought by education and training, they may not be obliged to sit through a program of learning. Another related benefit is that competency-based programs lend themselves to a self-paced learning model. Because of the explicit statement of outcomes in competency texts, learners can use the texts (or materials based on them) to gauge their own progress and take control of their learning.

A second major benefit claimed for CBE is that the education, training and assessment for specific jobs are highly relevant, because the competency texts represent current realities of the occupation. In contrast, other kinds of curriculum—for instance, those based on disciplinary knowledge or other traditional knowledge bases—have no guaranteed alignment to the target occupations and may not be subject to regular review. The "industry" focus of education and training can be assured when programs, delivery and assessment systematically reference competency texts (Smith and Keating 2003).

A third major benefit of CBE is that competency written statements/texts are the reference point for design and development of programs, delivery and assessment (Harris et al. 1995). A competency-based approach thus facilitates a high degree of alignment between the various elements of an education and training system, ensuring that the experience of learners is integrated and that different trainers and assessors are working from a similar knowledge base.

Finally, a competency-based approach fosters a modular approach to the design of training materials (Cornford 1997). The analysis and representation of competent work usually yields multiple competencies. The resulting modular structure means that implementation of competency-based programs in particular organizations or industries can be highly efficient, delivering only the education and training perceived to be necessary for local needs.

CBE thus promises a range of efficiencies that cannot be easily obtained from traditional approaches to education and training. For organizations, industries and systems, the benefits increase over time because generally shorter bouts of training are required for potentially more relevant outcomes. Individual learners also benefit by being recognized for their existing expertise and by receiving education and training only when necessary to achieve competence.

On the other hand, a range of challenges has been articulated by researchers and stakeholders. An assumption of CBE is that competence is a determinate state that learners either have or do not have. The CBE approach expresses this assumption in static descriptions of competent performance that do not suggest different levels of expertise. But many accounts of skilled performance differentiate several levels of attainment. For example, the famous Dreyfus and Dreyfus (1986) model of expertise describes a continuum of performance from novice through to expert level. They provide generic descriptions for each level of performance. In contrast, CBE descriptions tend to describe only a single level of performance.

Other researchers have shown that expert performance has a fundamentally "social" dimension. For example, in Lave and Wenger's (1991) theory of situated learning (discussed earlier in the chapter), becoming a capable practitioner is always something done in the context of a "social practice" and expertise is something that has a social as well as individual basis. From this perspective it is questionable whether it is legitimate to focus only on the competence of individuals. Rather, it is reasonable to talk about the competence of a team as much as of an individual. Therefore, the individual focus of CBE is challenged by some contemporary learning theory.

Another assumption of CBE is that crucial features of competent work can be identified using appropriate techniques or that they can be communicated by professionals. However, there is a body of work suggesting that expertise possesses embodied and tacit dimensions, and that language cannot adequately express these kinds of knowledge (Dunne 1993). This body of work, which stretches back to Aristotle and through to more recent contributions by Wittgenstein (1958), Heidegger (2010), Merleau-Ponty (2012), Polanyi (1958), Dreyfus and Dreyfus (1986) and Schön (1984), indicates that technical-rational means may not be appropriate for capturing expertise nor textual forms for representing it.

A related challenge is that the number and variety of competencies that usually issue from analysis of occupations, and the modularization of education and training stemming from competency texts, may have the effect of fragmenting occupational roles (Buchanan et al. 2009). Modularization allows development of competence in an occupation to proceed one competency at a time, but according to some researchers, this approach makes it difficult for learners to develop a sense of the whole occupation and role. Subtle, underlying understandings or broader knowledge structures could be lost to the learning process—for example, if they escape the identification and representation processes or if they cannot be expressed within the format of competency texts.

Another challenge concerns the interpretation and translation processes that are a significant and inescapable feature of competency-based education and training. Designers, developers, trainers, teachers, assessors, learners, auditors and other stakeholders all need to read and understand the competency texts they work with. Theory of interpretation or "hermeneutics" suggests that there is always slippage between the intentions of writers of competency texts and the understandings derived by readers (Schmidt 2006). Divergence between intentions and implementations in a competency-based system is also supported by communications theory (which always anticipates "noise" that distorts messages between transmitters and receivers) and curriculum theory (which distinguishes between intended, taught and experienced curriculum and anticipates considerable differences between these in practice). Indeed, behavioral objectives theorists are among the few who claim that an instructional intention can be transmitted with little loss via appropriately constructed texts (Hodge 2014).

Conclusion

Despite some powerful challenges to the rationale and practice of CBE, the introduction of a competency-based system exposes and questions traditional assumptions about occupations, education and training, and represents an undeniable force for renewal. The introduction of CBE is generally a troubling and exciting event for occupations, organizations and systems, and presents a unique opportunity to pose fundamental questions about practices and purposes.

This chapter has examined essential features of CBE. CBE is characterized by a focus on competence, competency texts and educational practices in accordance with the texts. These features can be summarized as follows:

> **Competence:** the ability to fully participate in a complex social practice, such as an aviation profession. Full participation requires skills, knowledge and attitudes relevant to that practice.
>
> **Competencies:** negotiated and agreed written statements (texts) that attempt to represent the ability to fully participate in a social practice. Competencies embody assumptions about:
> - the nature of competence
> - the number of discrete written statements (texts) required to represent competence
> - the most appropriate language for representing competence.
>
> **Competency-based education (CBE):** instructional design, training and assessment that systematically references written competencies.

These statements serve as a descriptive summary of assumptions, processes and practices shared across different implementations of CBE. In the next chapter, the focus moves to the emergence of CBE within the aviation industry. Like regulators in other industries and sectors around the world, the International Civil Aviation Organization (ICAO) has responded to the contemporary trend to orient training to competence. We will see that occupations within the aviation industry have reacted in their own ways, both to the ICAO's new direction and to other influences that transmit the concepts of CBE.

Chapter 2

CBE from Above:
The ICAO and Underlying Regulations

Questions Considered in this Chapter:

- What is the International Civil Aviation Organization (ICAO)?
- How does the ICAO influence training and assessment in aviation?
- Why is the aviation industry adopting CBE and how did this happen?
- How are CBE regulations created and structured?
- Which aviation professional groups are adopting CBE?

Short History of Aviation Regulations

Aviation is a relatively young industry that took off after the Wright Brothers' first flight in 1903. In 1909, only six years after the first flight, the French voiced concerns about aircraft entering France—or overflying it—that had not originated from within its borders. These issues created discussions around rules pertaining to countries' rights with regard to who could depart from, fly over or land in a particular country, and these discussions became a precursor for the "freedoms of the air" rules currently in place. World War I certainly slowed the consolidation of international aviation rules and regulations. However, using some of the dialogue occurring from 1909 (and before), a convention was drafted in 1919 at the Paris Peace Conference. In 1922 it was signed and ratified as the Paris Convention, and one of its recommendations was the formation of the International Commission of Air Navigation (ICAN).

About 20 years later it became clear that the aviation industry was being restrained by increasing logistical and political obstacles. As the fundamental purpose of the aviation industry was worldwide transportation, its success was dependent upon international cooperation. In November 1944, 52 States (in regulatory terminology, the term *States* refers to countries) met in Chicago to debate a wide range of aviation regulatory issues. The result of these negotiations was the Convention on International Civil Aviation, also called the Chicago Convention, which was ratified by 26 States in April 1947. Currently, of the 195 countries in the world, 191 States adhere to the Convention.

The Chicago Convention consists of four parts: (1) Air Navigation, (2) International Civil Aviation Organization (ICAO), (3) International Air Transport and (4) Final Provisions. Born out of part 2 of the Chicago Convention, the

ICAO is a United Nations special organization headquartered in Montreal, Quebec, Canada. The ICAO is the international forum for civil aviation. Today, the goal of the ICAO is to achieve sustainable, safe and secure aviation through cooperation among member States. One point that must be made is that the ICAO, while attempting to always implement best practice and practices that have been studied through scientific investigation, will in the end implement *what can be agreed upon.*

The ICAO is made up of two bodies:

1. *Assembly*—a group composed of representatives from all countries who have signed on to the Convention. All States have an equal right to representation and voting.
2. *Council*—a permanent body responsible to the Assembly, which is composed of 36 States elected to three-year terms.

The Chicago Convention contains 96 articles outlining the restrictions and privileges of the States, along with details on International Standards and Recommended Practices (SARPs). A Standard is defined as follows:

> Any specification for physical characteristics, configuration, material, performance, personnel or procedure, the uniform application of which is recognized as necessary for the safety or regularity of international air navigation and to which Contracting States will conform in accordance with the Convention; in the event of impossibility of compliance, notification to the Council is compulsory under Article 38.

A Recommended Practice is defined as follows:

> Any specification for physical characteristics, configuration, material, performance, personnel or procedure, the uniform application of which is recognized as desirable in the interest of safety, regularity or efficiency of international air navigation, and to which Contracting States will endeavour to conform in accordance with the Convention.

Annexes are important additions to the original Convention. Annexes are added to the Convention when updates are necessary to address issues that were not originally considered. For example, Annex 16 addresses environmental protection and Annex 17 covers security, both of which were less of a concern in 1944 than they are now.

The very first Annex added to the Convention was on Personnel Licensing—for example, licensing of air traffic controllers and engineers. Annex 1, which was proposed by the ICAO Council in April 1948 and adopted by the Assembly later that year, contains SARPs that serve as minimum standards for personnel licensing. Annex 1 is of particular importance in the discussion of competency-based education and will be referenced throughout this chapter and Part II of this book. It is interesting to note that someone reading the 1948 version of Annex 1 would notice that the Standards are very similar to those of today. Although some changes have been made, the basic structure of the Standards has remained the same (Dow and Defalque 2013).

The Challenge

Back in 1919, long before Annex 1 was drafted, some of the earliest standards were created for pilots, and pilots needed to pass a series of tests to achieve the "Pilot's Flying Certificate for Flying Machines Used for the Purposes of Public Transport." The tests aligned quite nicely with the skills required for real-world flying—as a single pilot. No minimum number of training hours was specified for the certificate. However, by 1920, the ink was barely dry on the Standard when States began implementing its recommendations and introduced a focus on experience, which was measured by a required number of hours (PIL8).[1] This focus on hours of experience has had a lasting impact on aviation training. To this day, pilots meticulously track every flight hour throughout their training and professional experience. A specific number of training hours are required to earn a license, and airlines require a few thousand flight hours before pilots are eligible for a job interview.

Today, the traditional hours-based licensing approach has a long training and regulatory history and an excellent industry safety record within a variety of aviation professions. Worldwide, a large amount of quality training is conducted under this methodology. However, the ICAO has recognized that global air transport is estimated to increase dramatically over the coming decades. For instance, a survey conducted by the ICAO "indicates a potential global shortage over a 20-year period of 160,000 pilots, 360,000 maintenance personnel and 40,000 air traffic controllers, for a grand total of 560,000 aviation professionals" (Elamiri 2013, 3). Hence, the challenge for the world aviation sector is to produce an adequate supply of qualified aviation professionals who are trained to support the future growth of the industry without compromising safety.

The fundamental question now is whether the training system should abandon the prescriptive focus on the number of hours and to instead focus on the knowledge, skills and attitudes required for professional competence. When a trainee must complete a prescribed number of hours in order to earn a license, there is no motivation or flexibility for the training industry to improve its efficiency through new approaches, such as simulation or new teaching technologies. The focus on hours limits training innovation (Dow and Defalque 2013).

---

1   PIL8: Airline Training Expert, Asia.

> In modern times, training companies may scoff at the idea of trying to reduce training hours, through more efficient approaches, as the more hours a trainee completes the more revenue the company receives. Most aviation training companies build their businesses on hours. ... That's the absurdity of the traditional pilot training approach. We can't focus the attention on the real world. We focus the attention on meeting the hours. (PIL3)[2]

The reason for the focus on hours is simple: hours are easy. Although not necessarily the best method of validating skills, hours provide a quantifiable and standardized language when discussing experience.

Measuring experience and training quality in terms of hours is also a convenient approach for regulators:

> Imagine for a moment that you are a regulator sitting in your office and a recommendation from the safety board comes across your desk, based on a recent accident investigation. If the recommendation is to take action associated with a particular license or rating, the easiest thing to do is to add a couple of hours [to the requirement]. Adding hours gives the appearance of having done something when, in fact, it probably does nothing to target the root of the problem. (PIL3)

A recent example of this approach is the United States' regulatory response to the Colgan Air Flight 3407 accident. The accident occurred on February 12, 2009, and all four crew members and 45 passengers were tragically killed, along with one person on the ground. The families of those killed in the accident lobbied the United States Congress for action, and the Federal Aviation Administration responded with the Airline Safety and Federal Aviation Administration Act of 2010. This response was seen as an example of "smoke and flames rulemaking" (Collins 2014, para. 2)—that is, a rapid knee-jerk response by regulators. The Act introduced new rules that required first officers to have an Airline Transport Pilot (ATP) certificate with at least 1,500 hours effective August 1, 2013. There is tremendous debate surrounding this issue, as many believe that additional hours do not necessarily result in better skills. Also, the additional hours requirement will make it more difficult for pilots to accumulate the required experience for airline employment, which results in regional airlines competing over the reduced supply of qualified pilots. Eventually this will result in higher airline fares for consumers without necessarily improving safety. Clearly, "adding a few hours of training" to address an issue may not target the root of the problem and may lead to new and unforeseen challenges. A shift in focus away from hours and towards competence is favored among many training professionals.

---

2   PIL3: ICAO (retired), North America.

Regulatory History of Competence: Reviewing Annex 1

It is interesting to recall that—at a time when much attention in the aviation community is given to competency-based education—the words *competence* and *competency* have been a part of ICAO provisions since their inception. The preamble of the first edition of Annex 1, adopted on September 15, 1948, stated that:

> The expression "license" used throughout these Standards and Recommended Practices (SARPs) for Personnel Licensing has the same meaning as the expressions "certificate of competency and license," "license or certificate" and "license" used in the Convention. (Interview 3)

This wording remains in Annex 1 today, along with the terms *degree of competency* and *maintain competency*. So, it is clear that the idea of competency is not new. It has been tied to aviation regulations since the early days of formalized international civil aviation, specifically for the granting of licenses to those who operate within and maintain the air transportation system.

It is also interesting to note that Articles 32 and 33 of Annex 1 specify that a person needs to meet a set of minimum requirements to become a license holder. The idea was that a person who exhibits a set of minimum requirements has demonstrated competence. Therefore, from a regulatory perspective, we can consider the terms *requirements* and *competency* to be related.

In the case of a pilot, competence is demonstrated by meeting a certain set of requirements. For example, Annex 1 states the following:

> 2.1.1.3 An applicant shall, before being issued with any pilot licence or rating, meet such requirements in respect of age, knowledge, experience, flight instruction, skill and medical fitness, as are specified for that licence or rating.
>
> 2.1.1.3.1 An applicant for any pilot licence or rating shall demonstrate, in a manner determined by the Licensing Authority, such requirements for knowledge and skill as are specified for that licence or rating.

The foresight of the writers of these SARPs in 1948 is impressive. They recognized that the approach to licensing and training aviation personnel would be likely to evolve as the industry learned more about instructional practices.

Amendments have been added to the document as aviation training has advanced. For example, the use of simulation or synthetic training devices in the assessment of skills was added to Annex 1 in the 6th edition in 1973. In the 1982 edition, a new idea was introduced in the section addressing maintenance personnel: competence was subdivided into competencies in procedures. It was not until July 2006 that a definition for the word *competency* appeared in the 10th edition of Annex 1 as a result of the ICAO adopting the multi-crew pilot license (MPL), which will be discussed in more detail in Chapter 4.

The following definitions are presented in Annex 1 (ICAO 2013):

- **Competency.** A combination of skills, knowledge and attitudes required to perform a task to the prescribed standard.
- **Competency unit.** A discrete function consisting of a number of competency elements.
- **Competency element.** An action that constitutes a task that has a triggering event and a terminating event that clearly defines its limits, and an observable outcome.
- **Performance criteria.** Simple, evaluative statements on the required outcome of the competency element and a description of the criteria used to judge whether the required level of performance has been achieved.
- **Threat management.** The process of detecting and responding to threats with countermeasures that reduce or eliminate the consequences of threats and mitigate the probability of errors or undesired aircraft states.
- **Training objective.** A clear statement that comprises three parts, i.e., the desired performance or what the trainee is expected to be able to do at the end of training (or at the end of a particular stage of training), the performance standard that must be attained to confirm the trainee's level of competence and the conditions under which the trainee will demonstrate competence.

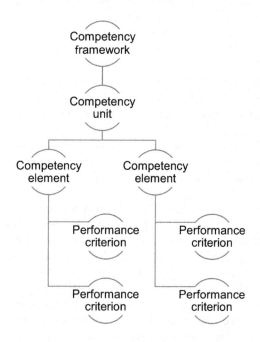

Figure 2.1   The Relationship between Competencies, Competency Units, Competency Elements and Performance Criteria

Competency units, competency elements and performance criteria are determined based on job and task analyses, and describe observable outcomes (ICAO 2006). For example, the competency frameworks for flight crew and maintenance personnel are based on the competency units shown in Table 2.1.

Table 2.1   Competency Frameworks for Flight Crew and Maintenance Personnel

Competency frameworks	Flight crew	Aircraft systems maintenance personnel	Aircraft structure maintenance personnel	Aircraft component maintenance personnel
Competency units	1. Apply threat and error management principles 2. Perform ground and pre-flight operation 3. Perform take-off 4. Perform climb 5. Perform cruise 6. Perform descent 7. Perform approach 8. Perform landing 9. Perform after-landing and post-flight operation	1. Perform fault isolation 2. Perform maintenance practices 3. Perform service 4. Remove component/assembly 5. Install component/assembly 6. Adjust 7. Test 8. Inspect 9. Check 10. Clean 11. Paint 12. Repair 13. Perform MEL and CDL/DDPB procedures	1. Perform aircraft structural repair inspection 2. Perform structural damage investigation, cleanup and aerodynamic smoothness check 3. Perform special process application 4. Perform metal rework/testing 5. Perform structural repair	1. Perform testing fault isolation 2. Perform disassembly 3. Clean 4. Perform inspection/check 5. Repair 6. Perform assembly 7. Perform storage

Source: International Civil Aviation Organization 2006.

The definitions of competence and competency should reflect the contemporary views on how practitioners progress from novice to expert, how levels of competency are dynamic depending on experience and context, and how cognitive ability and skills are called upon to support high-reliability organizations involved in the air transportation system.

Moving Forward with CBE and Regulations

Frontline aviation professionals such as pilots, air traffic controllers, maintenance personnel, ground crew and cabin crew, just to name a few, typically work in teams. Because the ICAO has focused on individual license holders, the regulations have been directed toward individual competencies. This can lead to an underestimation of the individual's competency as a team member within a specific context. Two members of a flight crew should demonstrate competencies that enhance that team's effectiveness.

So how do competencies fit into the human performance/factors world? With emerging technologies, and the considerations of human–automation interface, it could be said that not only do individuals have competencies, but teams and systems also have unique competencies. The way we define competencies needs to take into account a systemic view and how humans fit into it.

An enormous amount of confusion currently exists within the industry around the definition of the term *competence*, and additional work will be required to clarify this issue. Some believe that the competencies defined by the ICAO are not competencies at all—they are tasks. Overall, "because of the prescriptive nature of global and national standards it is very hard to combine the hours-based methodology and the real-world focused methodology" of CBE (PIL3). Regulators will have to work with the aviation industry to further validate the real-world focused methodology and to show that it is superior to the hours-based approach.

Part I

References

Bernstein, B. 2000. *Pedagogy, Symbolic Control, and Identity: Theory, Critique, Research*. Lanham, MD: Rowman & Littlefield.

Bloom, B.S. 1968. Learning for mastery. *Evaluation Comment* 1(2): 1–11.

Boyatzis, R.E. 1982. *The Competent Manager: A Model for Effective Performance*. New York: Wiley.

Boyatzis, R.E. 2007. Competencies in the 21st century. *Journal of Management Development* 27(1): 5–12.

Brockmann, M., L. Clarke, P. Méhaut and C. Winch. 2008. Competence-based vocational education and training (VET): The cases of England and France in a European perspective. *Vocations and Occupations* 1: 227–44.

Buchanan, J., S. Yu, S. Marginson and L. Wheelahan. 2009. *Education, Work and Economic Renewal: An Issues Paper Prepared for the Australian Education Union*. Sydney: Workplace Research Centre, University of Sydney.

Carroll, J.B. 1963. A model of school learning. *Teachers College Record* 64: 723–33.

Chomsky, N. 1965. *Aspects of the Theory of Syntax*. Cambridge, MA: MIT Press.

Collins, R. 2014, March 28. *A Double Tragedy: Colgan Air Flight 3407*. Blog post. http://airfactsjournal.com/2014/03/double-tragedy-colgan-air-flight-3407/. Accessed March 21, 2015.

Cornford, I.R. 1997. Ensuring effective learning from modular courses: A cognitive psychology-skill learning perspective. *Journal of Vocational Education & Training* 49(2): 237–51.

Dow, J. and H. Defalque. 2013. The MPL – A systems approach to pilot training gains acceptance by States and industry. *ICAO Training Report* 3(2): 31–3.

Dreyfus, H.L. and Dreyfus, S.E. 1986. *Mind over Machine: The Power of Human Intuition and Expertise in the Era of the Computer*. New York: The Free Press.

Dunne, J. 1993. *Back to the Rough Ground: 'Phronesis' and 'Techne' in Modern Philosophy and in Aristotle*. Notre Dame, IN: University of Notre Dame Press.

Elam, S. 1971. *Performance-Based Teacher Education: What is the State of the Art?* Washington, DC: American Association of Colleges of Teacher Education.

Elamiri, M. 2013. Plotting a path towards sustainable air transportation. *ICAO Training Report* 3(1): 3.

Garfinkel, H. 1967. *Studies in Ethnomethodology*. New York: Prentice-Hall Inc.

Glaser, R. 1963. Instructional technology and the measurement of learning outcomes: Some questions. In *Criterion-Referenced Measurement*, ed. W.J. Popham, 5–16. Englewood Cliffs, NJ: Educational Technology.

Gonczi, A., P. Hager and L. Oliver. 1990. *Establishing Competency-Based Standards in the Professions.* Canberra, ACT: National Office of Overseas Skills Recognition.

Harris, R., H. Guthrie, B. Hobart and D. Lundberg. 1995. *Competency-Based Education and Training.* South Yarra, VIC: Macmillan.

Hayes, J.L. 1979. A new look at managerial competence: The AMA model of worthy performance. *Management Review* 68(11): 2–3.

Heidegger, M. 2010. *Being and Time.* New York: State University of New York Press.

Hodge, S. 2007. The origins of competency-based training. *Australian Journal of Adult Learning* 47(2): 179–209.

Hodge, S. 2014. *Interpreting Competencies in Australian Vocational Education and Training: Practices and Issues.* Adelaide, SA: National Centre for Vocational Education Research.

Hymes, D.H. 1966. Two types of linguistic relativity. In *Sociolinguistics*, ed. W. Bright, 114–58. The Hague: Mouton.

International Civil Aviation Organization. 2006. *Doc 9869 Procedures for Air Navigation Services: Training.* Montreal, QC: ICAO.

International Civil Aviation Organization. 2013. *Doc 9995 Manual of Evidence-Based Training.* Montreal, QC: ICAO.

Joyner, C.W. 1995. The DACUM technique and competency-based education. In *Challenge and Opportunity: Canada's Community Colleges at the Crossroads*, ed. J. Dennison, 243–55. Vancouver, BC: UBC Press.

Kaslow, N.J., N.J. Rubin, M.J. Bebeau, I.W. Leigh, J.W. Lichtenberg, P.D. Nelson, S.M. Portnoy and I.L. Smith. 2007. Guiding principles and recommendations for the assessment of competence. *Professional Psychology: Research and Practice* 38(5): 441–51.

Kirwan, B. and L.K. Ainsworth. 1992. *A Guide to Task Analysis.* London: Taylor & Francis.

Kornhauser, A.W. 1922. A plan of apprentice training. *Journal of Personnel Research* 1(5): 215–30.

Lave, J. and E. Wenger. 1991. *Situated Learning: Legitimate Peripheral Participation.* Cambridge, UK: Cambridge University Press.

Leung, W.-C. 2002. Competency based medical training: Review. *British Medical Journal* 325: 693–5.

Lévi-Strauss, C. 1963. *Structural Anthropology.* London: Penguin Books.

Mager, R. 1962. *Preparing Instructional Objectives.* Belmont, CA: Lake.

McClelland, D.C. 1973. Testing for competence rather than "intelligence." *American Psychologist* 28(1): 1–14.

McLagan, P. 1980. Competency models. *Training and Development Journal* 34(12): 22–6.

Merleau-Ponty, M. 2012. *The Phenomenology of Perception.* Milton Park, UK: Routledge.

Patil, A. and G. Codner. 2007. Accreditation of engineering education: Review, observations and proposal for global accreditation. *European Journal of Engineering Education* 32(6): 639–51.

Piaget, J. 1952. *The Origins of Intelligence in Children.* New York: International Universities Press.

Polanyi, M. 1958. *Personal Knowledge. Towards a Post-Critical Philosophy.* London: Routledge.

Schmidt, L.K. 2006. *Understanding Hermeneutics.* Stocksfield, UK: Acumen.

Schön, D.A. 1984. *The Reflective Practitioner. How Professionals Think in Action.* New York: Basic Books.

Scott Tilley, D.D. 2008. Competency in nursing: A concept analysis. *Journal of Continuing Education in Nursing* 39(2): 58–64.

Smith, E. and J. Keating. 2003. *From Training Reform to Training Packages.* South Melbourne, VIC: Cengage Learning.

Talbot, M. 2004. Monkey see, monkey do: A critique of the competency model in graduate medical education. *Medical Education* 38: 587–92.

Taylor, F. 1904/1964. *The Principles of Scientific Management.* New York: Harper.

White, R.W. 1959. Motivation reconsidered: The concept of competence. *Psychological Review* 66(5): 297–333.

Wittgenstein, L. 1958. *Philosophical Investigations.* Oxford, UK: Basil Blackwell Ltd.

PART II
Aviation Professional Training

Introduction to Part II
Aviation Professional Training

Questions Considered in Part II

The following chapters discuss training considerations for four aviation professional groups—air traffic control, pilots, cabin crew, and maintenance engineers:

- What are the responsibilities of each of these professional groups?
- How were these professionals trained historically? How is contemporary training conducted?
- How is competency-based education (CBE) being incorporated into their professional curricula?

Introduction

As discussed in Part I of this book, CBE is being implemented into a variety of professional areas in aviation. Yet, the meaning of CBE varies widely among professionals:

> Competency-based [education] is more of a buzzword right now. … There's, unfortunately in our industry, a habit of taking a buzzword and turning it into something it isn't. Taking a term and mashing it together with whatever ideas or agendas people are pushing—there's no definitive understanding of what this term is. Yet, it is very difficult to change mindsets once they have caught on. (CC2)[1]

Most aviation training experts are intimately familiar with how professionals learn their trade, maintain skills throughout a career and acquire new expertise to move to greater levels of responsibility within their profession. Yet, the training conducted within related professional groups may be entirely foreign. To better understand CBE in aviation, it is important to understand how traditional training is conducted within professional disciplines, how CBE began in each role and what CBE looks like in present-day training. This understanding is crucial before one can consider how CBE might reshape future training practices.

This section of the book explores professional training practices of air traffic control, pilots, cabin crew and maintenance engineers, along with narratives from

---

1   CC2: Cabin Training Expert, North America.

training experts in each area. An effort has been made to remain neutral and avoid issues of rivalry that sometimes occur among professional groups in aviation. In the words of one air traffic controller, "What's the diplomatic way to say this? There is a mild irritation when there seems to be inferred that we can learn a lot from the pilots, because we see our job as fundamentally different" (ATC1).[2] The authors acknowledge that each of these professions have rich histories and practices that could not be captured within a single chapter. The intent of this part of the book is to highlight aspects of these professions that are relevant to an understanding of CBE.

The professional groups will be discussed in the following order:

Chapter 3:   Air Traffic Control.
Chapter 4:   Pilots.
Chapter 5:   Cabin Crew.
Chapter 6:   Maintenance Engineers.

---

2   ATC1: Air Traffic Control Training Expert, Africa & Europe.

Air Traffic Control and
Competency-Based Education

Brief History of Air Traffic Control (ATC)

There is no doubt that an efficient and safe air traffic control (ATC) system is a fundamental component of modern aviation. ATC began in the 1920s, when the Aeronautics Branch of the U.S. Department of Commerce issued uniform field rules in ATC. In 1946, an inspection of the ATC practices in the United States resulted in the provisional International Civil Aviation Organization (ICAO) aligning international standards with those of the United States. This can be considered the beginning of unified ATC services on a global scale.

Although radar was used in World War II as a means of detecting enemy aircraft, it was not until the 1960s that radar was integrated into civil air traffic systems. Later, individual radar sites became increasingly accurate, with a range of around 160 nautical miles, enabling air traffic controllers to safely and efficiently manage traffic flow into larger airports. In non-radar environments, the implementation of new navigation equipment on jet aircraft in the 1960s, such as distance measuring equipment (DME), began to change the way that ATC interacted with and separated aircraft.

Modern ATC Procedure and Environment

The main job of the air traffic controller is to prevent collisions between aircraft in the air, and when they are on the ground to prevent collisions between aircraft and obstructions. Furthermore, the controller is required to manage the aircraft in such a way that there is an expeditious and orderly flow of air traffic.

In general, modern air traffic controllers are able to separate aircraft from each other and guide aircraft to avoid terrain in three different ways: (1) visual reference, (2) procedural separation and (3) surveillance radar.

With *visual* reference, the controller is typically working at an aerodrome and is able to see aircraft as they approach the airfield. By directly observing the positions of the aircraft and knowing their intentions, the controller separates the aircraft by passing instructions relating to how they should route, what they should avoid and/or in which order they should proceed. For example: "Cessna 152 ABC, do you have the Piper Chieftain in your 1 o'clock position in sight?",

"Affirm, Cessna 152 ABC", "Roger, Cessna 152 ABC, position yourself number 2 behind that aircraft and report turning base-leg."

The second mode of separation is that of *procedural* separation. Here the controller will use separation rules, along with *dead reckoning* principles to separate aircraft. The most frequently used type of procedural separation is vertical separation, where the controller ensures that aircraft in level flight have at least 1,000 feet between them (with greater separation at higher altitudes). Other forms of procedural control include time separation, whereby one aircraft is instructed to cross a specific point at a specified time, and the second aircraft (at the same flight level) is instructed to cross the same point 10 minutes later.

The final method is that of *surveillance*. When providing separation using surveillance equipment, the amount of space needed between aircraft can be significantly reduced by comparison with procedural control. Surveillance of aircraft is typically achieved through the use of radar equipment, however newer surveillance equipment using satellite technologies is becoming commonplace. Surveillance equipment enables a controller to "see" all aircraft on a screen, which in turn allows the controller to separate the aircraft by ensuring that they keep a minimum distance away from each other. They do this by issuing instructions on what headings and tracks to fly and then monitoring the aircraft and adjusting instructions based on the aircraft's actual position.

ATC Professional Positions

In addition to there being different basic practices associated with how controllers separate aircraft, these practices are utilized by controllers to provide three main services: (1) aerodrome control (tower), (2) approach control and (3) area control [please note that the terminology used to describe these positions varies by geographic region]. For each of these services, a license and separate *rating* would need to be earned by a controller. The rating that is earned also depends on whether the service is provided with or without the use of surveillance equipment (for example, radar).

Aerodrome control service is located in a control tower. An aerodrome controller assists in the safe and efficient departure and arrival of aircraft at airports and ensures that aircraft flying in the vicinity of the aerodrome are separated from other aircraft. The required rating has two parts: the first is for those who separate aircraft flying with only visual reference to the ground and other aircraft, and the second is for those who separate those that are flying using the aircraft navigational instruments rather than visual references. However, in most cases, the aerodrome controller needs to be qualified to separate both types of flights. Note that there can be operational positions that include ground and air—which again can be assisted by radar facilities.

The second type of service is *approach control*. This service allows a controller to facilitate the departure and arrival of aircraft from one or more aerodromes and transitioning aircraft to and from area controllers. A controller is issued a rating that allows this service to be provided either with or without the use of surveillance radar.

The last type of service is that of *area control (also called en-route control)*. Here the controller facilitates the safe and efficient separation of aircraft throughout the cruise portion of the flight. These ratings recognize whether the service is being provided using procedural control or with the aid of surveillance radar.

A final consideration is the controller's *validation*. A controller requires a validation for each section of airspace they work within. For example:

> I can do air traffic control because I have a license. I can do aerodrome control because that's my rating. And I can do it in [my area]. But I can only do it in [my area]. If I wanted to go to [another area], I don't have to get a new rating because I'm already an aerodrome controller, but I do have to get a new validation, which is [for the new area]. (ATC4)[1]

During taxi and takeoff an aircraft is in touch with *aerodrome control*. Once an aircraft has departed an aerodrome, aerodrome control usually pass to *approach control*—although pilots normally refer to them as "departure"—who will facilitate the safe and efficient departure of the aircraft from the aerodrome area. The approach controller then hands the responsibility of that aircraft's separation to *area control*. Area control has responsibility for the aircraft until it begins its approach into its destination airport, where it will hand off to that approach controller and lastly an aerodrome controller from the destination, who guides the aircraft down to landing and taxi.

Traditional ATC Training

Although the ICAO produced standards for ATC licenses and their associated rating in the 1950s, regional differences existed in early ATC training. One air traffic controller who had experienced a 1960s-era curriculum recalled:

> I don't know the best way to describe it, quite loose training objectives that you could write into it whatever you wanted. That's essentially what I was trained off, a lot of theory. The practical application at the academic level was fairly minimal, very much being driven by the fact that we didn't have the technology we have nowadays to simulate the air traffic environment. So because the technology wasn't great, at the initial training level, you had a very small amount of practical training. The vast majority of the practical training was done when you went to the unit where you were going to work. The unit training itself was really practical, with no simulation lead-in. There was very little other than

---

1 ATC4: Air Traffic Control Training Expert, Southeast Asia.

understanding the letters of agreement between the unit you were working in and the adjacent units. *My lasting memory is you would do your training in the academic environment, and your first day on the job somebody was bound to say to you, know everything you learnt at the college? You can now forget it because now I'm going to teach you how to **really** do the job.* (ATC1, emphasis added)[2]

A major evolution in ATC training occurred around the late 1990s when task-based training was introduced. *Task-based training* is associated with an instructional design process that identifies job elements that compose the work of a controller and creates training syllabi specific to those identified elements. This explicitly aligns classroom instruction to real-world practice before a trainee begins on-the-job training (ATC1). A task-based approach is very similar to competency-based education (CBE), which is expanded upon later in this chapter, and can be considered an early form of CBE in ATC training. In modern training, simulation has facilitated the combination of theory and practical teaching through scenario-based instruction.

ATC Training Footprint: An Example

During the very early stages of initial training (referred to as basic training), the training footprints of aerodrome control (tower), approach and area control are very similar. After basic training is completed, specializations into the different ratings or combinations of ratings can be undertaken. For brevity, the training of area controllers will be presented as an example [please note that training differences occur regionally].

Initial training refers to a combination of classroom and simulation-based training, and can be understood as occurring in two phases:

Basic training: theoretical and practical training that is designed to impart fundamental knowledge and practical skills related to basic operational procedures. It allows trainees to progress to specialized ATC training.

Rating training (training in the rating discipline): theoretical and practical training designed to impart knowledge and practical skills related to a specific rating (Eurocontrol 2015).

During classroom training, which lasts approximately six weeks, students undertake face-to-face instruction—although e-learning is increasingly common—on a variety of topics such as the following:

- basic knowledge (ICAO, concepts of compliance)
- air law

---

2   ATC1: Air Traffic Control Training Expert, Africa and Europe.

- rules of the air
- Manual of Air Standards (e.g., separation requirements)
- inflight emergency response and search and rescue
- communication and standard phraseology
- aerodromes and other landing surfaces
- facilitation
- security and unlawful interference
- navigation
- meteorology
- principles of flight
- flight instruments
- aircraft radio and electronic aids
- conduct of flight operations
- aircraft type knowledge
- aviation occupational health and safety
- surveillance control.

Upon completion of classroom theory, learners begin simulator-based training (sometimes referred to as an "emulator," as they may not fully replicate an ATC center). The focus of this stage is to combine procedural (non-radar) and surveillance (radar) controlling, with stages that include the following:

- directed traffic information (e.g., low-level traffic, outside controlled air space)
- procedural control
- surveillance control
- combined operations (i.e., everything together).

In our example, the simulator-based training is approximately 40 to 50 weeks in duration. Simulation training is structured in a way that allows students to build up specific skills. As a student becomes increasingly comfortable with normal controlling, the traffic scenarios increase in complexity and unusual situations, and degraded systems are introduced. For example, initially:

> There are no equipment failures taking place. So we see that they are competent to perform in a very normal environment. And then the last learning unit is completely dedicated to integrating this standard performance while being able to deal with emergencies and unusual situations. So the philosophy behind it is first know how to do this job in an uninterrupted undisturbed way, and once you're able to do that, only then start bringing in these emergency situations. So that, in fact, during that last learning unit it's about integrating [dealing with emergencies and degraded situations] into a competence that you already have. (ATC1)

Upon completion of initial training, learners progress to on-the-job training within an ATC unit. During on-the-job training, new air traffic controllers will have an instructor "plugged in" via headset to assist and monitor their actions. This will enable the trainee to gain practice as a controller while being assisted and mentored by an experienced air traffic controller. Over the course of years, the trainee will develop expertise in managing traffic in a variety of sectors in order to work within a particular "group of sectors." At this point he or she will be known [in the region used in this example] as a "full performance controller" or FPC.

However, as in most professions, there are those who criticize the traditional approach to training. One criticism of this training method is that it is error focused:

> It was focused on the errors, not on what you did well. It was what you didn't do so well. And, I suppose, because they didn't have a competency-based system [with] clearly defined criteria ... then the feedback was always about what you didn't do. It wasn't bigger picture stuff. ... But also they didn't really have a picture anymore of what an air traffic controller did ... at the end, what they needed to look like. What were the important things? (CC3)[3]

To expand upon traditional training, in the 1990s ATC authorities in Australia began considering the impact of a variety of non-technical attributes on trainee performance: "We looked at not only just ... the knowledge and the skills to execute the task, but also those underlying values. So you know, their attitudes and beliefs, their background, and their culture ... and we're looking at fostering teamwork" (ATC3).[4]

ATC instructors in Australia implemented an interesting training method to allow instructors to assess competence, which they call a "prompting hierarchy" (ATC3). This approach defines four levels of prompting that an instructor would deliver to a student during a training exercise:

1. **Questioning:** "Is there something you need to do?" If this does not trigger the correct response from a trainee, the instructor moves to the next level.
2. **Suggesting:** "I would be looking in the northeastern corner for something to do up there." If this still does not lead to a correct action from the trainee, the instructor moves to the next level.
3. **Directing:** "What are you doing about those two aircraft, there? You need to do something, now."
4. **Intervening:** The instructor takes over for the trainee.

The prompting hierarchy is used to quantitatively measure how a trainee is progressing toward competence. If level 3 (directing) or level 4 (intervening) occurs during the final weeks of training, the trainee is not likely to be recommended for their proficiency check:

---

3   CC3: International Organization Representative, North America.
4   ATC3: Air Traffic Control Training Expert, Southeast Asia.

We would be expecting a controller, as they got to the end of their training, to be able to work the airspace, do all the required tasks, achieve separation, et cetera, with minimal if not no input from the on-the-job training instructor. So we'd expect there to be no questioning or suggesting. ... That prompting hierarchy is purely a quantitative measure at the end of training to say "okay, yes, we believe that this controller has progressed towards autonomy and they can do the job by themselves." (ATC3)

Competency-Based Education in ATC

Why is ATC training around the globe moving toward a competency-based approach? As one training expert explains:

What an air traffic controller did was missing from our students' or our trainees' vision, and we had no clear criteria [of] what made an air traffic controller. ... We needed to have a competency-based system. We needed some clear criteria on how we would know our students were competent to be air traffic controllers. (ATC4)

What is CBE for ATC? An explanation is presented by a senior training expert:

I would say that competency-based training is when the training—the instructors and anybody who's involved in that training—knows and understands from the very beginning what the end result needs to be. ... And then the route to that end result, to competence, can vary from person to person, depending on the knowledge and the skills that they already have. ... There's some important features or aspects that are present in competency-based training. The first one is that competence, that performance, it must be an integrated performance. So, integrating all of the competencies. It's not good enough to just train on single competencies and be able to tick off that single competencies have been achieved. ... You must have evidence that performance is consistent. So here we kind of move away from this one-shot examination and we say that you need to see over a series of simulator sessions, or a series of operational shifts, that that person is able to provide that integrated performance. You can't do those as a one-shot. (ATC1)

When Did CBE for ATC Begin?

It is important to point out that the evolution of CBE for air traffic controllers and pilots seemed to advance in parallel (around the same time, yet not directly linked together). Interestingly, there were several groups working to identify ATC competencies around the globe. Intended as an introduction rather than a comprehensive list, the following three examples will be discussed: (1) Australian Case, (2) New Zealand Case and (3) Netherlands Case. Next, the upcoming ICAO revision to the Procedures for Air Navigation Services – Training (PANS-TRG, Doc 9868) will be discussed, which is expected to come into effect in November 2016.

Australian Case: A Collaboration between Academia and Government

In the mid-1990s, the University of Queensland was engaged to remodel the performance assessment reports of Australian controllers. Representatives went around to each ATC location and observed controllers doing their job. They also did interviews and spoke to operational controllers, check and standardization staff, and managers. The result of this work was a performance assessment tool with three parts:

1. Antecedent determinants: what people bring to their role (skills, knowledge, experience).
2. Performance parts: communications, scanning, system interaction and contextual performance.
3. External factors: pilot actions, weather, unusual situations.

> What we'd like to move to for an assessment type process is more about, "okay … you got a good outcome this time, but it wasn't executed in a way which will give you that consistent outcome all the time." … So on our current model we'd say "oh well, they're not projecting and planning very well." But in actual fact they might have had a great plan but when they implemented it, it may have been the implementation that was actually the problem, rather than the planning. So our current model doesn't actually lead us to go back and assess what went wrong from that point of view. It's just purely an outcome-based type model, … we're finding that that is somewhat flawed in an investigation-type process. (ATC3)

This system also generated a competence model that includes four categories:

1. Cognitive competencies: planning, prioritizing and projecting
2. Communication: phraseology, coordination, etc.
3. System interaction: controller pilot data link, automatic dependent surveillance systems, etc.
4. Contextual assessment: completed by managers who assess teamwork, company values, work ethic and related items (this item does not directly impact licensing). (ATC3)

Although the Australian approach isn't considered fully competency-based:

> It's probably a bit of a blend because it doesn't really go into competent, not yet competent. What they did, I think, when they designed it was they [said to themselves] … we want a competency-based model because … in the mid-90s that's what CASA [the Civil Aviation Safety Authority] was moving to and that's

what they were moving towards. But they also wanted, within that competency-based model, a facility to identify expertise. So we said "okay, well this was the minimum standard to be competent," but you know, an expert would actually deliver the service or they would perform that competency far greater, or at a far greater standard than what the minimum level of competency was. So it is a blended type approach, it's competency-based training for the [mid-level performer] ... and then for anywhere where they make the gauge [closer to] expert type performance, then that was really modelled around being able to identify people that maybe do the job a little better than what somebody else does. (ATC3)

New Zealand Case: A Private Company Approach

An air traffic control service provider in New Zealand began working on their own internal competency model in 2001. They had considered using the model developed in Australia, although one training expert's opinion was that "it was just too heavy. It was too much detail, there were too many numbers, and it just wouldn't work in our system" (ATC4).

Therefore, they began the lengthy process of writing their own set of competencies. This was accomplished by gathering examiners and training specialists and asking, "What makes a good air traffic controller?" (CC3). Writing competencies included the following process:

> Once we had all these buckets [general categories], and with all the things that were in the buckets (like all the skills and all the knowledge that was in each of the buckets) then we just gave each bucket a name. So, situational awareness for a competency turned into "Apply situational awareness as the basis for informed decision making." So that was the competency. And then we just defined the criteria under it. But that went through, obviously, a lot of discussion ... defining, getting it right down, getting more people consulting, more opinions coming in, fine tuning it, and also establishing our new assessment regime. ... The competencies had to be clear, transparent, and not overwhelming, and then when we developed the assessment system for the competencies, that also had to be quite streamlined. (ATC4)

The end result of this approach was a list of technical and non-technical competencies; the last two on the list are the non-technical competencies:

1. situational awareness
2. managing traffic
3. conflict resolution
4. communication
5. coordination
6. effective use of equipment
7. observe and disseminate flight and weather information

8. administration (being up to date with changes and self-briefing)
9. self-manage performance (recognizing errors and self-correcting)
10. interpersonal relationships (with pilots, other controllers and peers). (ATC4)

Netherlands Case

A piece of foundational work that had a major impact in moving ATC toward CBE came from the Netherlands in 2006 (Oprins et al. 2006). The premise of this work was that training that focused on a learner's competence would be more effective, would lead to more informed pass–fail decisions and could increase the training output of competent controllers. To identify ATC competence, 12 controllers conducted "competence workshops" in which they drafted a set of competencies and then rank-ordered the competencies in terms of job relevance. The result was the identification of the following ATC competencies:

1. situation awareness (SA)
2. decisiveness
3. dealing with unexpected situations
4. workload management
5. conflict solving
6. multitasking
7. prioritizing
8. coordination and communication
9. flexible planning
10. leadership
11. teamwork ability
12. perseverance. (Oprins et al. 2006)

Then, each competence was given 8 to 12 behavioral markers. For example, the behavioral markers for "flexible planning" include the following: "1. Makes a plan, executes the plan, and adapts the plan to (changed) circumstances, 2. Reverts to standard procedures if necessary, 3. Adapts his/her own plan to requirements and wishes of others …" (Oprins et al. 2006, 304).

This listing of competencies and behavioral markers was then evolved into a performance model that divided the list into processes (e.g., actions, information processing) and outcomes (Oprins and Schuver 2003). It was clear in the ATC performance model that information processing held a dominant role in the work of controllers.

The final phase of Oprins and colleagues' work was the creation of an assessment system to determine when controllers had achieved competence. This required the drafting of performance criteria for competencies. For example, the performance criteria for workload management include the following:

- adapts his/her work tempo to traffic load
- minimizes his/her own workload as much as possible
- stays calm, also during hectic moments (Oprins et al. 2006, 307).

However, other elements of Oprins et al.'s (2006) assessment system included considering the learner's phase (level of expertise), incorporating continual assessments and performance tests, and requiring assessors to use web-based tools to store and standardize data. This model is presented and discussed further in Chapter 11.

As a senior instructor explains, this model has been adapted and used within some ATC training schemes:

> The model was defined [for] air traffic control instructors to apply it. … So they divided up the competencies into thinking processes, actions, and how you handle traffic. The behavior you saw, and then some individual aspects, so under these thinking processes we were talking about situation awareness, attention and workload management, planning and decision-making. Whereas your actions, it would be communication (very clearly observable), coordination (very clearly observable), how you operate your system (observable). In terms of behavior it was how was the teamwork and cooperation? What is your professional attitude towards what you are doing? … And then there was this individual aspect as well where we looked at how they cope with stress, personal attitudes towards learning, criticism, learning from mistakes, that kind of thing. (ATC1)

ICAO Guidance

Based upon the rich body of work that has already been discussed, it is logical to expect that the ICAO would be working on a global competency framework for air traffic controllers (ATCOs) and air traffic safety electronics personnel (ATSEP). This framework has been drafted and is expected to come into effect in November 2016. The goal of establishing a global competency framework was to create a regulatory structure that (1) allows the profession to incorporate modern training and learning technologies; (2) defines competencies for all aviation activities that impact safety, in order to allow the free flow of professionals around the globe through agreed-upon standards and assessment practices; and (3) supports the demand for new air traffic management procedures for those who work with increasingly complex technologies (ICAO n.d.).

The ICAO competency framework for ATCOs and ATSEP was developed by the Next Generation of Aviation Professionals (NGAP) Task Force, a group of air traffic management industry experts who discussed and reached consensus regarding the competencies required for these professions. The ATCOs and ATSEP competency frameworks are listed in Table 3.1.

Table 3.1   Listing of Competency Units from the ICAO's ATCO and ATSEP Competency Frameworks

	ATCO Competency Framework	ATSEP Competency Framework
Competency Units and *Definition*	Situation Awareness: *Comprehend the current operational situation and anticipate future events*	Engineering: *Collaborate in developing, modifying and integrating systems, networks and equipment*
	Traffic and Capacity Management: *Ensure a safe, orderly and efficient traffic flow and provide essential information on environment and potentially hazardous situations*	Situation Awareness: *Comprehend the current status of the ATM [air traffic management] system and anticipate future events*
	Separation and Conflict Resolution: *Manage potential traffic conflicts and maintain separation*	Service Provision: *Ensure availability and reliability of [communications, navigation, surveillance] CNS/ATM systems and capabilities*
	Communication: *Communicate effectively in all operational situations*	Coordination: *Manage coordination between personnel in operational positions and with other affected stakeholders*
	Coordination: *Manage coordination between personnel in operational positions and with other affected stakeholders*	Management of Nonroutine Situations: *Detect and respond to emergency and unusual situations related to the ATC operation and/or CNS/ATM systems and capabilities*
	Management of Nonroutine Situations: *Detect and respond to emergency and unusual situations related to aircraft operations and manage degraded modes of ATS [air traffic service] operation*	Problem Solving and Decision Making: *Find and implement solutions for identified hazards and associated risks*
	Problem Solving and Decision Making: *Find and implement solutions for identified hazards and associated risks*	Self-Management and Continuous Learning: *Demonstrate personal attributes that improve performance and maintain an active involvement in self-learning and self-development*
	Self-Management and Continuous Development: *Demonstrate personal attributes that improve performance and maintain an active involvement in self-learning and self-development*	Workload Management: *Use available resources to prioritize and perform tasks in an efficient and timely manner*
	Workload Management: *Use available resources to prioritize and perform tasks in an efficient and timely manner*	Teamwork: *Operate as a team member*
	Teamwork: *Operate as a team member*	Communication: *Communicate effectively in all situations*

Source: ICAO n.d.

Industry Perspectives of CBE

From the perspective of ATC industry trainers, a competency-based approach to training and assessment offers advantages over traditional approaches. For example, the ability to *assess learners at different phases* was particularly useful, because "it allowed the instructors a very clear reference to what they should be able to do at 200 hours. Because, if you didn't provide that standard, they were always measuring against the end point standard" (ATC1).

CBE also allows for instructors to be more *objective in their assessments* of trainees:

> Instructors get really close to their students, or their trainees, and so there's a little bit of halo effect ... so it wasn't always, I don't think, fair or equitable. When we went and flagged competency, it became a lot more objective, rather than subjective. I mean, in assessment there's still subjectivity, you can't get away from that. But it became, they had better tools in order to make objective decisions around their students' performance. (ATC4)

In addition, CBE is also a good tool to determine *why people make mistakes*, rather than just diagnose the mistake itself, and create problem-focused training:

> In the past [training] was ... about correcting each mistake, it wasn't understanding the bigger picture of "Well, actually all of these mistakes are fitting into this bucket." ... Concentrate on the learning, not on the error, because the error would just show up as something different a different day, and the instructors would go "Oh, they can't do it, they're making all these errors." But actually we hadn't addressed the errors, and the student didn't know how to fix them, because they didn't understand it, they didn't know why they were making them. (ATC4)

One trainer gave an example of a trainee losing separation in a simulated environment and emphasized the importance of understanding why it happened:

> Was it because they didn't have appropriate situational awareness? Was it because they were unable to regulate their workload? Was it because they just simply haven't learnt the letters of agreement and have not therefore been able to apply it? So the model, the competency-based model almost allowed you to consider a whole range of, of areas where there might be a deficiency, instead of just honing in on one mistake: the person lost separation. And the real trick, the instructors who are really good at this quite often would be able to work out where the deficiency was. ... [But] it only worked when the instructors really knew, really understood the model, and understood that if you were going to mark somebody ... down on a particular competency, you needed to reflect and you needed to appreciate that there was some other competencies that were just so interconnected that you couldn't then mark them as satisfactory in other areas. (ATC1)

Yet this leads directly into one of the challenges of CBE, which is the *re-education of instructors*:

> I suppose one of the things that we needed to change, and we needed to do a lot of work on, was reframing our instructors. So teaching them how to use competencies, but also reframing their language, focusing on the bigger picture ... if [trainees are] making a bunch of mistakes, where is that, is it in their situational awareness? ... So [instructors are] better able to unpack what is happening, and they are more effective in their training and their feedback in coaching the trainee in order for them to be able to improve, and how the trainee understands it and what they need to do. So if you don't train your instructors appropriately, then to use the competencies, then that's a negative, you don't really have a competency system at the end of the day. (ATC4)

Another drawback of this approach is the *lengthy documentation* associated with CBE and, as one expert suggests:

> There is no way that the instructor is going to be able to sit with this guide in front of him and look at a screen as well and assess and instruct. You have to know those 35 pages really intimately. And if you're not instructing continuously, you're not going to know the model to that level of depth. So the downside is that when somebody is instructing or assessing, if they're not 100% familiar with all of these descriptions, then their assessment is going to be less accurate. (ATC1)

Within ATC training, another CBE issue is associated with the *duration of training*. CBE allows advanced learners to progress through the training curriculum more quickly, but the current structure does not slow down training to give extra time to those who are progressing slowly. Some instructors intentionally do not mention this principle (that, theoretically, CBE could take as long or as short a time as required to achieve competence). They reason that in practical training environments, it would not be realistic to invest the additional time and money into a trainee who is not progressing within a specified period of time (ATC1). This is logical from a logistical resource-based perspective, but it may represent an area for additional research in the future.

Another challenge is associated with how instructors *assess "invisible" cognitive skills*:

> From the instructing point of view the biggest difficulty we have, which I'm sure everybody has, is it's very easy to see the actions that are being performed. It's impossible, well it's not impossible but it's unbelievably difficult, to try and work out whether the cognitive skills have been achieved. So of course in our model we have a whole load of cognitive skills that we require [of] the controller. We of course have planning skills, we have the decision-making

skills, we have the whole gambit there. And of course our problem is well, well how do you know that they've actually acquired these skills and it's associated with competency? (ATC1)

Lastly, the analysis, design and development of CBE can be *prohibitively complex* to many ATC trainers. "The people who are designing the training are not necessarily qualified to do ISD [instructional systems design]. They are quite often controllers who went into instructing, who went into then training in a college, and just through experience have fallen into designing the training" (ATC1).

Conclusion

Overall, it is clear that air traffic control has a rich training history as well as a proud culture. It is interesting to note that the origin of CBE within this profession was internal, growing from controller-focused initiatives in several regions including Australia, New Zealand and the Netherlands. As CBE increases in popularity, it is important to continue to explore the true advantages and challenges of this approach and best practices associated with its use in ATC training.

Chapter 4
Pilots and Competency-Based Education

The piloting profession is very diverse. A pilot's mission varies significantly depending upon the type of operation he or she is involved in, such as crop dusting, instructing, helicopter flight, military operations, regional airlines or transoceanic international flights on the new Airbus A380. Clearly, pilots face a range of operational challenges and their training must prepare them to handle both routine and unexpected situations.

A Brief History of Pilot Training

Early pilot training was focused on two areas: classroom and practical. Classroom instruction dispensed knowledge about aircraft structures, flight controls, aerodynamics, engines and rules of the air. The practical application was learning to fly, or the "hands-and-feet" skills. For example, pilots would undertake lessons in taxi, takeoff, circuits and landings. Early pilot training was generally conducted solo; pilots would go out by themselves to conduct exercises that had been discussed with the instructor. Unfortunately, this approach led to many accidents and deaths in the early years of aviation training (Moroney and Moroney 1999). For example, 10% of United Kingdom pilot losses during World War II were in training accidents (PIL8).[1]

A critic of this approach was Major Robert Smith-Barry, a decorated World War I pilot. He wrote basic training principles and suggested that pilots should practice dangerous maneuvers and be taught how to recover. He standardized instructor training and introduced dual-seat aircraft, which enabled instructors to demonstrate maneuvers. Smith-Barry set up the Central Flying School at Gosport, England, and began conducting training with many of the same instructional principles that are prevalent in the training systems of today (Mavin and Murray 2010).

Advancements that took place between World Wars I and II led to additional instruction for pilots. For example, pilots were now able to fly aircraft over longer distances, which necessitated instruction on navigational techniques. Another development was cockpit instrumentation that enabled pilots to fly in clouds without visual reference to the ground (instrument meteorological conditions [IMC]).

During World War II, aviation saw some of its greatest advancements. Pilots were flying aircraft with a co-pilot more often, rather than as a single pilot. Many flights incorporated a team of crew members that could include navigators,

---

1 PIL8: Airline Training Expert, Asia.

engineers, radio operators and gunners. Additionally, aircraft were now at higher altitudes, and pilots were being exposed to hypoxia and cold weather, which could lead to hypothermia. This brought about the first aviation medicine training requirements for pilots (later called "human factors" or "non-technical" training).

One of the most important training developments was the introduction of the flight simulator, with the earliest simulators introduced around 1909. The flight simulator was able to replicate, with varying degrees of fidelity, the flight characteristics of an aircraft. For a longer discussion on the impact of simulation on training, please see Chapter 10.

Overview of Traditional Licensing and Training

As it is based on ICAO standards and recommended practices, the training path of pilots is fairly consistent around the globe. Pilots initially begin training as student pilots. Students who possess no previous aviation experience are referred to as *ab initio*, which is a Latin term meaning "from the beginning." Students complete a combination of classroom, in-aircraft and usually simulator lessons until, after about 12 to 18 months, they earn a commercial pilot license (CPL). It should be noted that a predetermined minimum number of aircraft flight hours is required to achieve a CPL. A CPL is endorsed as fixed wing (airplanes), rotary wing (helicopter) or balloon. With a CPL, pilots are restricted to flying smaller aircraft as pilot in command (PIC), although they can act as co-pilot on larger aircraft. As with their air traffic control and engineering colleagues, ratings (such as an instructor or an instrument rating) can be added to a pilot's license.

As pilots gain experience (as measured by number of flight hours completed), they move into roles of increasing responsibility. As an example, pilots who wish to fly for an airline can be tested for the Airline Transport Pilot License (ATPL) after completing 1,500 hours of flight experience (with early hours completed in the training environment and the remainder collected during real-world operations in the early stages of their professional pilot career). The ATPL allows the pilot to take on greater responsibility as captain of a high-capacity aircraft, such as the Boeing 767. At this level, pilots will have to achieve *type ratings* on each additional aircraft they want to fly. For instance, a qualified Boeing 767 pilot would have to achieve an additional type rating before piloting an Airbus A320. Because of the complexity of individual aircraft types, and their associated operating procedures, airlines restrict pilots to flying only one type of aircraft at a time.

Industry Challenges: Projected Pilot Shortage

One of the biggest challenges facing the aviation industry today is the proposed shortage of aviation professionals. Global air transport has doubled every 15 years since 1977, and the number of commercially operated aircraft worldwide is expected to increase from 62,000 to 152,000 by the year 2030, when 60 million flights (carrying 6 billion passengers) are forecast annually. Also, the Asia Pacific region has grown from being the third largest to the largest air transport market in the world in just over a decade (Benjamin 2013). This rapid and sustained expansion has led to an especially high need for pilots in the Asia Pacific region. The International Civil Aviation Organization (ICAO) predicts a potential global shortage of 160,000 pilots over the next 20 years, with approximately 40% of demand coming from the Asia Pacific region (Elamiri 2013, Mayerowitz and Koenig 2015). The challenge for the world pilot training sector is to ensure that an adequate supply of qualified aviation professionals are trained to support future industry growth, without compromising safety. This has led the pilot training industry to rethink traditional practices and search out opportunities to train pilots more quickly and also to a higher quality standard.

The Shift toward Competency-Based Education (CBE)

Given the recognized need for the industry to produce more qualified pilots, an informal ICAO meeting was held in Madrid in 2000 to explore the future of licensing and training standards. This led to the creation of the ICAO Flight Crew Licensing and Training Panel (FCLTP) in 2002 (Dow and Defalque 2013). It was at this time that the term *competency-based* first began to be used; however, it was noted that "none of us understood the term but it opened the door to a systems approach in pilot training, which was well-understood" (PIL3).[2]

The FCLTP learned that the focus on hours as a training requirement was a particular problem. For example, in Europe, 70 hours of pilot in command (PIC) time was required for the commercial pilot license. This requirement seemed unreasonable for pilots being trained for a first officer position in a two-crew environment where they would not act as PIC. Although regulations require 1,500 hours of flight experience for an ATPL, the activities conducted and challenges faced by pilots within the course of those flight hours will vary significantly, and therefore 1,500 flight hours does not directly correlate to a level of piloting competence. A pathway was needed around traditional hours-based licensing regulations, and the FCLTP began exploring alternatives (Dow and Defalque 2013).

An early goal of the FCLTP was to create a *new pathway* to achieving a pilot license, rather than to change the existing licensing structure of the CPL and ATPL. The reason for this was that the FCLTP could not change the existing licensing standard and put 191 States out of compliance! The goal was to create a new, optional alternative licensing approach. The FCLTP developed the multi-crew pilot license (MPL) to fill this need (PIL3). However, the MPL was not necessarily based on scientifically identified best practices: "the way you develop

2   PIL3: ICAO (retired), North America.

international standards is not through a scientific process. You deal with it through what people can agree to do" (PIL3).

Although the MPL standards are similar to those of traditional licenses, with specified requirements for skill, knowledge, age, medical fitness and experience, there are important differences (Dow and Defalque 2013). The MPL was the first competency-based pilot license and it focused more on an analysis of piloting practices in a *multi-crew* flight deck than traditional training that focused on single-pilot operations. The multi-crew focus is an important point because, before the FCLTP meetings, flight schools would complain about the single-pilot focus of the commercial license. For example, during the pilot's flight test, if the candidate asked the examiner to act as a crew—such as find information or throw a switch—they would be failed (PIL8).[3]

However, an MPL licensing structure would only be practical if there was a regulatory structure that allowed its adoption. In the course of the FCLTP's work, an interesting discovery was made. A short provision buried deep within the regulations stated that:

> A licensing authority may approve a training program for a private pilot license (PPL), commercial pilot license (CPL) or instrument rating (IR) that allows an **alternative means of compliance** with the experience requirements established by Annex 1 provided that the approved training organization (ATO) demonstrates, to the satisfaction of the authority, that the training provides a level of *competency* at least equivalent to that provided by the minimum experience requirements for personnel not receiving such approved training. (ICAO Annex 1, quoted by PIL3; emphasis added by authors)

This tiny provision hidden at the back of Annex 1 had a major impact, as it opened the door to an alternative means of regulatory compliance. In many ways, this can be viewed as a blank page. If a training organization creates a very powerful methodology, there is an open door to discuss anything with the regulator (PIL3).

Moving Forward: The Foundation of Competence within the MPL

Before the FCLTP could create the MPL as a competency-based license, they needed to determine the content of the training program. What does a competent pilot look like? How can competence be identified, documented, taught, assessed and eventually enforced? The FCLTP considered a few perspectives that provided insight into pilot competence: (1) the advanced qualification program (AQP), (2) crew resource management (CRM) and non-technical skills, and (3) threat and error management (TEM). The following sections explore each of these perspectives and discuss how they provide data to demystify pilot competence.

---

3   PIL8: Airline Training Expert, Asia.

Advanced Qualification Program (AQP)

One approach explored by the FCLTP was that pilot competencies were linked to the tasks completed during flight. The Advanced Qualification Program (AQP), first developed in the United States, was identified as a source of data that could inform this process. Please note that AQP focuses exclusively on airline-based training and is not currently applied to *ab initio* or other sectors (such as helicopter or corporate flight training). In the mid-1990s it was clear that "airlines weren't interested in understanding training. Airline training [was] simply maneuver repetition until you [got] it right or until you satisfied the regulatory requirement" (PIL8).[4]

AQP is a voluntary alternate pathway for traditional pilot training regulations. It allows airlines to use different training approaches as long as the company can justify that it results in equivalent or improved safety (Longridge 1997). AQP is accomplished through a task analysis that leads to the identification of training objectives for all levels of pilot training, which is an instructional systems design (ISD) process (PIL8). A key component of AQP is that air carriers must create and integrate a strategy for continual data collection to diagnose pilot skill and a method of implementing modified training materials based on this data (Longridge 1997).

Some distinguishing features of AQP include the following:

* voluntary
* innovative training and qualification concepts (as long as proficiency and safety can be demonstrated)
* proficiency-based qualification—with each air carrier developing its own terminal proficiency objectives (TPOs) within an ISD process (Longridge 1997).

The mandatory requirements of AQP are as follows:

* aircraft specific
* include indoctrination, qualification and recurrent training for every duty position
* incorporate, as much as possible, training in a full cockpit crew environment
* integrate training and evaluation of CRM
* provide AQP-specific training for instructors (including both training and evaluation strategies)
* collect performance data on students, instructors and evaluators, and conduct airline internal analyses on this information to refine training
* use advanced flight training equipment, including full flight simulators (Longridge 1997).

---

4   PIL8: Airline Training Expert, Asia.

AQP gained popularity within the United States through the 1990s. Meanwhile, major European airlines:

> were looking at what was happening under AQP in the States and the supposed cost savings (such as line checks every two years and only needing to come to the simulator once a year) and they put a marker down in the regulations saying they would like to take advantage. They were saying there was a potential competitive disadvantage between major airlines on either side of the Atlantic. (PIL8)

As a result, an approach similar to AQP, called the Alternative Training and Qualification Program (ATQP), was introduced in Europe in 2006. However, whereas AQP applies to all aspects of airline pilot training, ATQP only applied to recurrent training because of the section of the regulations that it is categorized within. This makes it administratively inconvenient to conduct training under two European regulatory systems: the traditional regulatory approach to training for indoctrination and qualification, and ATQP for recurrent training. However, a large number of the main carriers in Europe are now operating ATQP (PIL8).

For AQP and ATQP to be effectively integrated, an understanding of instructional systems design is crucial (PIL8). To integrate AQP in the United States, the Federal Aviation Administration (FAA) appointed a human factors scientist to administer the program (Weitzel and Lehrer 1992). This has been an advantage for the U.S. integration of AQP, compared with Europe, where "there is no expert sitting in the center with a background in ISD directing [the program]" (PIL8). In much of the aviation industry worldwide, AQP is poorly understood. "The background is just not there in [ISD] training—certainly in aviation flight training—people don't understand what you're talking about, therefore they don't see the benefits or understand the need" (PIL8).

The AQP and ATQP programs essentially consist of continual data collection, which allows the training curriculum to adapt to organizational and individual needs (though the ATQP system is much weaker than AQP in this regard) (PIL8).[5] This is accomplished by capturing numerical grades for technical tasks and then assigning associated letter designations (called reason codes) to capture any non-technical factors that contributed to the grade (Farrow 2010). This generated a significant amount of data that was useful to the FCLTP in understanding pilot competency. The AQP system can also be viewed as a training approach that is very similar to CBE, for several reasons: it allows carriers to be creative in customizing their training curriculum (rather than following prescriptive guidelines that may not perfectly fit the needs of their organization), it uses a task analysis to generate curriculum and it is based on texts (written statements) from the analysis to guide trainers and assessors (PIL8). AQP can result in cost savings for the organization, but it also has the potential to enhance professional skills beyond the qualification levels defined in regulations (Weitzel and Lehrer 1992).

---

5 PIL8: Airline Training Expert, Asia.

The difficulty with AQP as a model for a competency framework is that it is task and procedure oriented and may not fully consider, integrate or assess non-technical skills. While AQP data is relevant in normal operations, some industry experts feel it is not as applicable when confronting unforeseen and novel problems that require creative thinking, which is a crucial consideration in the definition of competence (PIL9).[6] In fact, the Air Line Pilots Association (ALPA) only supports AQP as long as a pilot cannot fail because of crew resource management (CRM) factors alone. The justification for this is that if CRM is very poor, it will eventually manifest itself as a failure on a technical task, with the CRM problem considered a contributing factor only (Farrow 2010). To understand the implications of non-technical skills on pilot competence, it will be helpful to review the evolution of CRM and threat and error management (TEM) training.

Non-Technical Skills

While some FCLTP panelists advocated the use of AQP to define competence, others held the perspective that pilot competencies were driven by non-technical skills. Crew resource management (CRM) was the first widely adopted system of non-technical pilot training. CRM training focuses on crew coordination and flight deck management (Helmreich and Foushee 2010). CRM is a significant departure from traditional training (before the mid-1970s) that focused on the psychomotor and cognitive *technical skills* required to safely maneuver an aircraft, but not the interpersonal, communication and task management *non-technical skills* that are developed in crew-based coordination training. The goal of CRM is to optimize the use of all available resources on the flight deck and for the crew to work effectively as a team—in contrast to the historic perspective, which often regarded the first officer as a secondary (less important) position and regarded skilled work as an *individual*, rather than *team*, accomplishment.

Following a time period with a number of high-profile human error accidents, CRM was introduced into the aviation industry in the 1970s. United Airlines, in the United States, was the first to integrate line oriented flight training (LOFT), which was scenario-based CRM training that engaged participants, allowed them to identify their unique managerial style and recurred through annual simulator-based sessions to refresh the concepts (Helmreich and Foushee 2010). Yet, the issue of human error on the flight deck remained a challenge in the industry. Some suggest errors had always been there, however, the advent of cockpit voice recorders and flight data recorders enabled a more holistic understanding of accidents. Nevertheless, from 1980 to 1989, more than 72% of the worldwide airline hull loss accidents were caused by avoidable pilot error (only a very slight improvement over the previous decade; Weitzel and Lehrer 1992). In more recent years LOFT has been renamed line operational simulations (LOS; Weitzel and Lehrer 1992).

---

6 PIL9: Instructional Designer, North America.

LOS, along with CRM, are incorporated within AQP programs. With this incorporation, experts are rethinking the separation between technical and non-technical training.

In parallel to the integration of AQP and ATQP into aviation training, an evolution of CRM was also developed in Europe. Beyond the use of CRM as a training domain, consideration of non-technical skills can also be used as a framework for assessing achievement through the use of behavioral markers to evaluate these skills (PIL8); NOTECHS (non-technical skills) is a behavioral marker framework created for assessing pilot CRM skills (Flin et al. 2003). Most airlines now use the NOTECHS framework as a default. The framework consists of four categories, which are subdivided into elements (see Table 4.1), which are then broken down into behavioral markers (see Table 4.2).

Table 4.1   Categories and Elements of the NOTECHS Framework

Category	Element
Cooperation	Team-building and maintaining standards Considering others Supporting others Conflict-solving
Leadership and managerial skills	Use of authority and assertiveness Providing and maintaining standards Planning and coordination Workload management
Situation awareness	Awareness of aircraft systems Awareness of external environment Awareness of time
Decision-making	Problem definition and diagnosis Option generation Risk assessment and option selection Outcome review

Source: Flin et al. 2003.

Table 4.2   Example of NOTECHS Behavioral Markers (Cooperation Category)

Element	Good Practice	Poor Practice
Team-building and maintaining standards	Establishes atmosphere for open communication	Blocks open communication
	Encourages inputs and feedback from others	Keeps barriers between crew members
	Does not compete with others	Competes with others

Source: Flin et al. 2003.

These behavioral clusters are each associated with a five-point scale and incorporated into the training assessment process. The NOTECHS framework also establishes five principles to ensure fair and objective assessments. This was deemed necessary because of the more subjective skill set required to evaluate non-technical compared with technical skills:

1. Assessments are based only on observable behavior.
2. For non-technical skills to be rated unacceptable, there must be a technical consequence (meaning the poor non-technical skill must have led to compromised flight safety).
3. The non-technical skills observed must be given an acceptable or unacceptable rating.
4. Unacceptable behaviors must be observed multiple times to conclude there is a problem.
5. Examiners must explain any unacceptable ratings (Flin et al. 2003).

Although the NOTECHS framework was a leap forward in the systematic incorporation of CRM into pilot training, it was challenging for some operators to integrate. "Whatever airlines were using to assess technical proficiency was in place and then we bolted behavioral markers onto the side of that as well" (PIL8).

Some organizations adapt the NOTECHS framework to meet the needs of their organization. Rather than use the 1–5 scale, after an exercise in the simulator they use the marker framework to guide the debrief discussions (PIL8). It is helpful to view the framework as an adaptive jumping-off point to begin the discussion of non-technical skills within an organization, rather than as a fixed set of guidelines that must be followed explicitly.

The challenge with using this model as the basis for a pilot competency framework is that it is difficult to assess NOTECHS objectively (although it must be appreciated that *all* assessment has some level of subjectivity). It is also challenging to regulate whether or not professionals are competent in these attributes. The added value of this model is that it ties competencies to human non-technical skills (PIL9).

Threat and Error Management

A third factor considered by the FCLTP when exploring pilot competency was threat and error management (TEM). As pilots, like all human beings, are subject to making mistakes, the aviation industry has adopted an error management approach. From this perspective, it is expected that a pilot will face threats during operations and that those may lead to errors. These mistakes are called *active failures.* Elements of risk, such as a slowly failing mechanical component, may not immediately impact the safety of a flight but are considered *latent conditions* that could eventually lead to an unsafe situation (Reason 1990, 1997). Active failures can be considered errors and latent conditions threats (Thomas 2004). TEM is a system used to identify and respond to internal or external influences that could reduce the level of safety in an operation (Helmreich et al. 1999).

TEM can be considered an evolved form of CRM training. Where CRM training focused on generic non-technical skills (such as communication, workload management, situation awareness and decision-making), TEM investigates the threats and errors most commonly encountered by crews within a specific aviation context and incorporates non-technical training interventions that are unique to that organization. This is accomplished by conducting a line operations safety audit (LOSA), which involves an expert observer pilot sitting in the cockpit jump seat, between and behind the captain and first officer, and collecting data on the threats pilots encounter and the errors made during normal operations (Helmreich et al. 1999, Klinect et al. 1999, Thomas 2004). This allows for the development of TEM training that is customized to the threats and errors most commonly encountered within a specific company.

It is important to note that there is not universal support for TEM, as some regard it only as a data collection tool (and that it is only through CRM training that safety improvements are made). In this sense "it's not 'evolved CRM' – it's subverted CRM. ... Also, when you look at error outcomes, pilots appear not to 'manage error' but rather construct revised operational plans to accommodate the new circumstances they are facing" (PIL8). Similarly, LOSAs are not perfect systems and can be considered "a pretty crude tool" (PIL8). They can reveal some interesting insights into how crews vary their performance according to the task demands of the port they are flying into, but they have several disadvantages: LOSAs are quite intrusive, crews may alter normal behavior because of the observer watching them and it is difficult and costly to create a data set large enough from which to draw strong conclusions (PIL8; Dekker 2003).

Overall Competency Approach of MPL

In the end, the MPL competency framework adopted an approach most closely aligned with the AQP model and analyzed "phases of flight" to produce competencies. *Competency units* were created to correspond to phases of flight, while *competency elements* correspond to high-level tasks within each phase of flight and *performance criteria* correspond to discrete evaluation points (see Figure 4.1).

Approved training organizations (ATOs) implementing the MPL have to customize their own training curriculum, based on the MPL competency framework, through a rigorous instructional systems design (ISD) process. This is required because the MPL competency framework is generic and does not suit the needs of specific airline contexts (see Chapter 8 for a discussion of ISD). ATOs must also detail a training blueprint that clearly links every teaching objective back to a specific competency (PIL9).

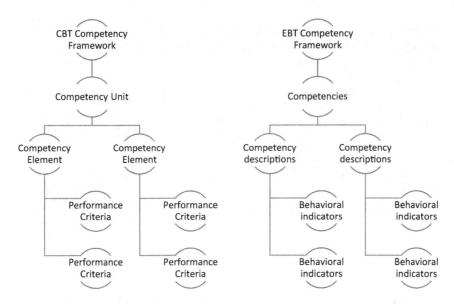

Figure 4.1   Example of Different Terminology used for Competency-Based Training and Evidence-Based Training

Interestingly, there is one competency in the MPL framework that cuts across all others, and it is associated with TEM. The competency requires pilots to recognize and manage threats and errors (Thomas 2004). One important point made during the deliberations of the FCLTP was that many conventional pilot training programs include modules *about* non-technical skills such as human factors, CRM and TEM, rather than training *how* to recognize and manage threats and errors in a real-world context. Within the MPL framework, ATOs *must* integrate TEM training while emulating the actual work that graduates will be doing: in an airline, in jet aircraft and in a multi-crew environment (PIL9).

After the FCLTP completed their work on the MPL competency framework, it was implemented in a few regulatory actions:

- new requirements for ATOs, developed and published in the Manual on the Approval of Training Organizations (Doc 9841) in 2004, and a second edition in 2012
- a new Procedure for Air Navigation Services (PANS) document (Training Doc 9869) in 2006, with chapters devoted to MPL implementation
- a State Letter in December 2007, requesting data on MPL implementation and training programs (Dow and Defalque 2013).

While the MPL is considered to be a potential solution for *producing* pilots more quickly, its original intent was to allow the development of training and licensing programs based on the expectations of how a pilot should *perform on the job*, rather than on a set of *prescriptive measures* (PIL9).

Currently, MPL is regarded as a safe and effective training option for airline pilot careers. As of January 2013, nearly 2,200 pilots had enrolled in MPL programs and 760 had graduated (Dow and Defalque 2013). However, MPL is actively used within only a few States because a very sophisticated and advanced regulator is required to implement this approach. Whereas licensing in terms of prescriptive hours is relatively straightforward, a competency-based approach can be complicated to oversee. Most States stick with traditional licensing approaches because they simply do not possess the structures to manage MPL (PIL3).

EBT versus CBE in Aviation Training

The structure established for the MPL competency framework (including competency units, elements and performance criteria) has been used for competency frameworks of all the other aviation professional groups, with one exception. The evidence-based training (EBT) used by the International Air Transport Association (IATA) Training and Qualification Initiative (ITQI) uses a completely different structure. The goal of EBT is to "develop a new paradigm for competency-based training and evaluation of airline pilots based on evidence" (ICAO 2013a, 2). The intention is to focus pilot recurrent training on key issues identified from data on normal operations, accidents and incidents. EBT has a heavy focus on the NOTECHS behavioral marker/standards systems. Therefore, EBT is regarded as a competency-based approach to recurrent training within flight simulation and training devices (FSTDs; ICAO 2013a).

Unfortunately, this can lead to some confusion within regulatory groups, as well as among professionals, around the differences between CBE and EBT. An easy way to remember the distinction between EBT and CBE is that EBT *exclusively* refers to airline pilot recurrent training, while CBE applies to all professional groups and experience levels. Adding to the confusion is that different structures and terminologies are used within CBE and EBT. For example, MPL and other competency frameworks follow a certain hierarchy: competency unit, competency elements and then performance criteria. In EBT, the competencies are organized in a different structure: core competencies and their set of behavioral indicators (ICAO 2013a; see Figure 4.2). The EBT structure uses the following definitions:

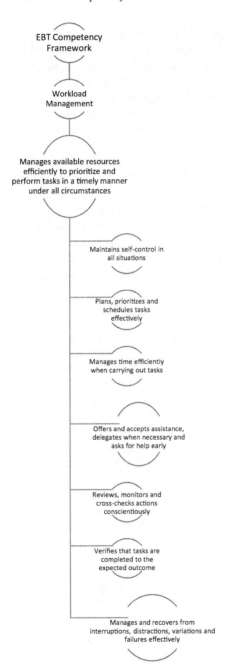

Figure 4.2   Sample EBT Competency Description and Behavioral Indicators

Source: ICAO 2013a.

- **Behavioral indicator:** an overt action performed or statement made by any flight crew member that indicates how the crew is handling the event.
- **Competency:** a combination of knowledge, skills and attitudes required to perform a task to the prescribed standard.
- **Competency-based training:** training and assessment that are characterized by a performance orientation, emphasis on standards of performance and their measurement and the development of training to the specified performance standards.
- **Core competencies:** a group of related behaviors, based on job requirements, which describe how to effectively perform a job and what proficient performance looks like. They include the name of the competency, a description and a list of behavioral indicators.
- **Evidence-based training (EBT):** training and assessment based on operational data that is characterized by developing and assessing the overall capability of a trainee across a range of core competencies rather than by measuring the performance in individual events or manoeuvers (ICAO 2013a).

With the introduction of EBT in ICAO provisions, an interesting question arose around exactly what competence and competencies are. The core competencies proposed in EBT are derived, with some adjustments, from technical and non-technical skills frameworks. The behavioral indicators underpinning each EBT competency were derived by polling the views of a large number of pilots, who were asked what constituted a "good pilot" (PIL9).[7] It could be argued that this approach to deriving behavioral indicators is impacted by the same strengths and weaknesses of the methodology described above for deriving competency frameworks. For example, one senior pilot stated, "I have read the EBT text including the competency. It looks great on paper until you have to actually use them. The competencies just don't work" (PIL7).[8] Another senior pilot commented:

> From a practical perspective, when you start looking at the amount of competencies that they put into that system, it's not a usable system from an airlines perspective. You just cannot write a … training system or an assessment system to look at and cover all those competencies. Once again, I don't think it targets the root cause. It's looking at phase of flight. And what the problem was with that particular phase of flight, rather than the holistic view. The good side of it is the work and the analysis they've done on where the problems exist worldwide, on certain types of aircraft, is very valuable in designing a matrix, training matrix so that you can actually look at the historic data and say, well,

---

7   PIL9: Instructional Designer, North America.
8   PIL7: Training Manager, Southeast Asia.

that's where everyone's saying the problem is, we actually agree with that. So it helps you build a matrix for your cyclic training system. (PIL5)[9]

However, it can also be argued that these opinions do not target the competency approach in itself, but rather are an indication that quality and effectiveness of competencies may vary. If "competencies 'don't work' then you have the wrong competencies!" (PIL8).[10]

Conclusion

Overall, it is clear that CBE has already had a significant impact on aviation training. Yet competency-based practices (such as AQP/ATQP) were impacting pilot training before formal competency-based regulations and licenses (such as MPL) were introduced. In considering how competencies are understood in present-day pilot training, a retired regulator summed it up nicely:

> You can see in the PANS [Procedures for Air Navigation Services] training document that what was really being asked was to implement a systematic approach to training. What we thought that meant was a systems approach to training. ... Then, this morphed into competencies because the commission had asked us to look at competencies. *I think we articulated a set of competencies that really were not competencies and the whole world is now dealing with this.* (PIL3)

---

9   PIL5: Training Manager, Southeast Asia.
10   PIL8: Airline Training Expert, Asia.

Chapter 5

Cabin Crew and
Competency-Based Education

Introduction

Cabin crew is often regarded as an occupation that facilitates glamorous globe-trotting adventures. However, it is clear that this profession has faced its share of challenges over the last 80 years. One issue has been the seemingly opposite roles of safety and customer service that cabin crews must balance. These roles will be discussed throughout this chapter, along with considerations related to licensing and the evolution of competency-based education (CBE) within the profession.

The realities of the job of cabin crew are more diverse than many aviation professionals appreciate, as illustrated by the events of September 11, 2001. One poignant example of a flight attendant carrying out her duties is that of Amy Sweeney. Sweeney was working on board American Airlines Flight 11 from Boston to Los Angeles, one of the planes that was hijacked and flown into the World Trade Center in New York City. Just an hour after speaking to her young daughter by phone, Sweeney demonstrated great bravery in flight when she called the company's flight service center "to report her airplane had been hijacked. She told the manager where the hijackers had been sitting, that she had been shown a bomb, and that two of her colleagues and a passenger had been stabbed" (Whitelegg 2007, 1).

A Brief History of Cabin Crew

Despite 1920s marketing slogans that touted air travel as being safer and generally better than traveling by land or sea, most passengers of that era knew that air travel was an unpleasant and risky experience. Passengers were packed into confined spaces, cabins were not insulated against the cold or engine noises, oil and fuel smells accompanied you during and after the flight, and the aircraft themselves were unable to fly around or over weather (which resulted in bumpy and uncomfortable conditions). Despite this, airline passenger travel was beginning to expand during the 1920s. For example, in the United States 9,000 passengers were carried in 1927; this number grew to more than 400,000 by 1930 (Barry 2007).

Cabin crews began to enter commercial aviation during the 1920s, when German and English airlines hired teenage "cabin boys." By the late 1920s, adult men were hired as stewards (Barry 2007). A woman named Ellen Church, a nurse and trained pilot (whose gender prohibited her from flying as a pilot in an airline), was the first advocate of women as cabin crew members. She organized a meeting with Boeing Air Transport (later to be called United Airlines) to suggest that nurses would be very good in the steward role. At the time, Stephen Stimpson of senior management reflected:

> it strikes me that there would be great psychological punch to have young women stewardesses or couriers, or whatever you want to call them. ... Also imagine the national publicity we could get from it, and tremendous effects it would have on the travelling public. ... Also imagine the value that they would be to us not only in the neater and nicer method of serving food and looking out for the passengers' welfare, but also in an emergency. ... it would be a mighty fine thing to have [nurses] available, *sub rosa*, if necessary either for air sickness or perhaps something worse. ... The average graduate nurse is a girl with some horse sense [who] has seen enough of men to not be inclined to chase them around the block at every opportunity. ... [Nurses would handle onboard clerical duties] better than the average young fellow. (Quoted in Barry 2007, 18–19)

Church's proposition to hire nurses as cabin crew gained popularity up until World War II. The war effort resulted in a mass deployment of qualified nurses into the military, requiring airlines to amend their hiring practices for cabin crew (Ball 2013). This led to the hiring of cabin crew with similar qualifications to those of the current day.

Unlike their other professional colleagues, cabin crews have experienced significant changes in their roles over the years. For example, during the 1930s, cabin crew would have been involved in loading bags, ensuring that all the seats were screwed down tightly and assisting with the refueling of the aircraft. The introduction of faster aircraft able to avoid weather and fly at higher altitudes made the position of cabin crew far less risky. By the 1950s, passengers had fewer concerns about being killed on the flight and more associated with lost luggage (Handy 2002). Food services were increasingly offered by airlines to bolster their marketing programs:

> a first-class meal might have included turtle soup served from a tureen, Chateaubriand carved seatside, and cherries jubilee. Steaks would be cooked to order—eggs, too, on breakfast flights. This is a world that, for obvious reasons, is even harder to conjure after September 11. (Handy 2002, para. 12)

As we move to the 1950s, the high hiring standards for cabin crew presented many obstacles for aspiring stewardesses. For example, in 1958, when jet travel was in its infancy, it was estimated that between 3% and 5% of applicants for stewardess positions were hired. At about the same time, Trans World Airlines boasted that it hired less than 3% of its applicants. For comparison, in 2006 Harvard University had an acceptance rate of 10.5% (Handy 2002).

Today, cabin crew are required to be trained in many areas, such as first aid, with some countries requiring all cabin crew to have a certified medical certificate prior to joining an airline. Cabin crew must demonstrate safety skills for emergencies such as cabin depressurization, as well as emergency evacuations, safety briefings for specific situations and fighting fires. Since September 11, 2001, cabin crew have become the last line of defense between the cabin and the flight deck. Yet, while shouldering all of these responsibilities, cabin crew are also at the whim of the airlines, which may realign their role for marketing purposes at any time.

The Fight for Recognition

Since the introduction of cabin crew in the 1920s, the most remarkable commonality that has occurred through the evolution of the profession is the fight for recognition of their work. It has been difficult, even today, to separate the safety role of cabin crew member from that of a marketing instrument of the airline. For example, cabin crew in some airlines are part of the service department, rather than operations.

Another challenge is associated with *unwritten* age restrictions for cabin crew. In the case of pilots, the "age 60 rule" has been hotly contested for decades. This mandatory retirement age for pilots was based on the recognized reduction of capacity associated with aging. However, with improved medical treatments and increased life expectancy, a pilot's age is no longer a valid measure of one's abilities (and is regarded as age discrimination). In contrast, consider how age has impacted the role of cabin crew over the years. In 1954 it was observed that:

> Because service is one of the chief areas of competition among the lines, the companies increasingly are stressing the importance of the girls. That same year American became the first airline to impose the mandatory retirement age of 32, thereby hoping to ensure its stewardess corps's ongoing pulchritude. (Handy 2002, para. 26)

The position evolved both in its role and how it was regarded within the aviation community as a profession. However, it was a slow evolution. In the United States, for example, following the enactment of the Civil Rights Act of 1964:

> federal courts in 1968 had struck down the rules forbidding marriage and forcing thirtysomethings into retirement. In 1970, restrictions against flight attendants' being pregnant were, under pressure, voluntarily withdrawn by many airlines. And in 1971, in the decision that would arguably do the most to put an end to the "coffee, tea, or me?" era, the Supreme Court ruled that airlines could not discriminate against men. (Handy 2002, para. 37)

One would assume that in modern times these issues would be dead and buried. However, the industry still lacks full transparency. For example, in 2005 an Australian state court ruled that Virgin Blue had discriminated against older women. The decision was based on the airline's insistence that potential cabin crew had to sing and dance as a way of demonstrating "Virgin Flair." Older women often did not get through stage one of the interview. Although the airline refuted allegations that its practices were discriminatory, it acknowledged that by the end of 2002 (it began operations in August 2000) it had not hired anyone over the age of 36 (Todd 2005).

Historical Training Programs

The historical approach to cabin crew emphasized attributes that seem odd from a modern-day perspective:

> stewardesses were told how to stand, how to walk, how to style their hair, how to make themselves up. Their "look" was as polished as the marble in a corporate lobby, and quality control was no joke: a woman who flew for TWA remembers that, aside from garden-variety infractions such as forgetting one's hat or getting caught smoking in uniform, stewardesses could be suspended if their milky complexions were darkened or freckled by too much time in the sun. A former Eastern Air Lines steward recalls being plucked from a flight for having a bruise on her leg, as if she had been a damaged piece of fruit blighting a grocery-store display. (Handy 2002, para. 7)

Likewise, the content of cabin crew training programs reflected the odd pairing of safety and service/marketing skills. For example:

> Having survived the initial winnowing—aside from multiple interviews, the screening process might have included I.Q. and psychological tests (if only the F.B.I. were as thorough)—potential stewardesses were dispatched to training centers for what was typically a six-week course of instruction. The facilities could be quite lavish: some had swimming pools and tennis courts; some were actually on the grounds of resorts. Given that their female charges had been selected for pheromonal impact, the schools had unique security issues. Most came equipped with curfews and guards. A few went further: the dorm-room balconies at Braniff's International Hostess Training College in Dallas had cage-like bars, allegedly because too many would-be suitors had tried to scale them. Perhaps for similar reasons, American's Stewardess College near Dallas was at one point surrounded by an eight-foot-high electrical fence. "One likes to think that the fence was to keep the intruders out, rather than the students in," notes *Wings of Excellence*, a semi-official history of American's flight-attendant corps. If you can conceive of a cross between Acapulco and a P.O.W. [prisoner of war] camp, that seems about right. (Handy 2002, para. 17)

As for the curriculum, it was generally divided between safety training and ladylike comportment, as indicated by the following points, taken from the outline for a 1964 lecture at United's training center:

Coat: 1. How to carry properly; 2. How to put on properly.

Together Look: 1. Coat always buttoned; 2. Wear gloves; 3. Carry everything on one side if possible; 4. Ways to carry purse; 5. How to carry gloves; 6. Scarf in summer raincoat.

Review: 1. Posture; 2. Standing; 3. Walking. (Handy 2002, para. 18)

Contemporary Training Programs

Nowadays, although there appears to be an increasing level of outsourcing, it seems that the majority of airlines conduct cabin crew training in-house. Training programs are typically between 5 and 12 weeks in duration. Early in training, students might complete medicals, go for uniform fittings and attend talks by senior executives. Classroom-based activities in the early weeks include introduction to aircraft, including concepts such as aerodynamics.

Upon completion of classes, cabin crew will be taken out to an actual aircraft to be familiarized with the flight deck and cabin. Cabin familiarization includes cabin configuration (first class/business and economy), basic door operation, interphone, galleys and emergency equipment positions. Students will then undergo more detailed classroom activities associated with emergency procedures, including drills for evacuations, ditching, smoke/fire, emergency equipment (such as the fire extinguisher and protective breathing equipment [PBE]). Training will then encompass service delivery for all flights (short- and long-haul) and all classes of passengers (e.g., economy versus business class).

Industry experts indicate that there are a range of training challenges associated with present-day cabin crew training:

As an overarching statement the issue [with cabin crew training] is always [that] training doesn't match with reality on-the-line. We train people to pass tests based on what the authority will be looking at. Airlines develop training to pass tests. So, let's say, you have a requirement to do a decompression. Every year cabin crew has to go through annual recurrent training. ... [For example, doing] decompression training every year and everyone gets a good grade. But then the airline has problems with inadvertent slide deployment. When you look at their data, because the regulator tells them they have to do decompression training every year, they do it, but that's not actually their training problem on-the-line. ... [Their training needs aren't] addressed in training because they are forced to do that same exercise over and over. (CC3)[1]

We're training people to meet regulations, not equipping people with what they need to manage issues on-the-line. We are doing it backwards, instead of saying "what do people need" and then reverse engineering training to match those competencies. ... How do we make cabin crew training more relevant in terms of reflecting problems on-the-line instead of just ticking boxes of what the regulator wants. (CC3)

The general consensus [associated with the problems in modern cabin training is that it] is always time and money. That's standard throughout the industry. From the pilot's side they get all the big toys all the time. Sometimes it's more difficult to acquire the funding to get cabin trainers and [e-learning]. The investment seems to be more difficult to come by ... for cabin crew throughout the industry. (CC2)[2]

To License or Not to License

An issue that is the focus of much debate around cabin crew training is licensing. Unlike other aviation professions, there is no international license required for cabin crew. Annex 1, discussed in Chapter 2, provides licensing provisions for pilots, mechanics and air traffic control but not for cabin crew. "To be frank, it's an historical issue. At the time, when they made those requirements, cabin crew weren't considered serious enough. Historically it wasn't something that was considered" (CC3). In addition:

The States [regulators], airlines, and unions push in different directions regarding licensing. The airlines don't want licenses because it means jumping through more hoops, the union does want it (you can say it's an issue of status and pay), I know in EASA one factor that pushed them toward the attestation was to increase work mobility within the European Union. (CC3)

In Europe, recent European Aviation Safety Agency (EASA) regulations have created a license for cabin crew that is called an *attestation*. "They didn't want to call it a license because that is a 'pilot-word'" (CC3). However, all cabin crew in States that follow European regulations must hold this attestation.

Under EASA's attestation program, candidates of at least 18 years of age undertake an approved curriculum that includes the following subjects:

---

1   CC3: International Organization Representative, North America.
2   CC2: Cabin Training Expert, North America.

- general theoretical knowledge of aviation and aviation regulations, covering all elements relevant to cabin crew
- communication: during training, emphasis shall be placed on the importance of effective communication between cabin crew and flight crew, including communication techniques, common language and terminology
- introductory course on human factors (HF) in aviation and crew resource management (CRM); this course shall be conducted by at least one cabin crew CRM instructor
- passenger handling and cabin surveillance
- aero-medical aspects and first aid
- dangerous goods in accordance with the applicable International Civil Aviation Organization (ICAO) Technical Instructions
- general security aspects in aviation
- fire and smoke training
- survival training. (Adapted from EASA 2012)

The European system uses a modular approach that identifies the elements of training that are standard across the profession (compared with those that are company specific). Standard elements include regulations, human factors training and general items associated with fixed components of the aircraft. This allows cabin crew to move between European airlines without starting training again from the beginning; instead, experienced cabin crew skip over standard training topics and just complete airline-specific curriculum. Therefore, the EASA attestation increases a crew member's mobility between airlines within the European Union and reduces training times (CC3).

In contrast, for cabin crew outside of Europe, training is specific to the operator they are working for. If an experienced cabin crew member quits one company and goes to work for another airline in the same country, he or she is required to complete training from the beginning (alongside inexperienced new hires).

The debate continues around whether cabin crew should be licensed on a global scale. The international union that represents cabin crew (the International Transport Workers' Federation [ITF]) is supportive of licensing and has made presentations at several ICAO Assembly meetings to request this change—so far unsuccessfully:

> The rationale as to why ICAO does not do it is because it's an administrative burden on the State. It's not specific against cabin crew, right now ICAO doesn't want to create requirements for any new license because it's a big burden for States to go through issuing all the licenses and then maintaining them and making sure they are current. (CC3)

Although some stakeholders desire the creation of cabin crew licenses, CBE may help the issue. As one expert points out:

for the cabin crew, if ICAO is developing this guidance [around CBE], even without licenses there are quasi-requirements for training. If through this approach you get competent cabin crew then what does the license add to it? That is the question. If you're training people to competency and they are competent ... is there an added benefit to having a piece of paper [license] to justify the administrative burden to issuing and maintain licenses? (CC3)

Competency-Based Cabin Crew Training

How Were Cabin Crew Competencies Written?

In 2014 the ICAO Cabin Crew Safety Training Manual was published as "guidance for a competency-based approach to cabin crew safety training so that cabin crew members could be proficient to perform their duties and responsibilities, and with the goal of establishing an international baseline for cabin crew competencies" (ICAO 2014, v). This work was accomplished by putting together the ICAO Cabin Safety Group, which was made up of approximately "50 members including regulators, airlines (covering all regions), IATA [International Air Transport Association] for industry representation, ITF for union representation, and aircraft manufacturers. From the get go, the group said, let's not talk about a license now. Let's talk about how we train crew to be competent" (CC3).

The ICAO Cabin Safety Group brainstormed cabin crew competencies throughout all phases of flight. They detailed what cabin crew would be doing throughout each phase, including emergency scenarios, and went through a process of reaching a consensus within the group to determine the high-level competencies:

> Cabin competency-based training flowed from the pilots, from an AQP [Advanced Qualification Program] perspective. We work on the same aircraft and we work as a team with the pilots. It really was just taking something good for the pilots and popping it into the cabin. At ICAO they developed the competency-based approach for the [Multi-Crew Pilot License] so ... for the cabin manual (it was an existing manual describing what you should have in training) ICAO decided to go with a competency-based approach because that is where the Organization is going in terms of philosophies for all the manuals. For the cabin competencies, ICAO based them on the pilot's model. ICAO took their framework, for example in normal operations competencies are based on phase of flight like a pilot, because pilots and cabin crew are on the same aircraft. Then ICAO looked at the pilot's framework and did a task analysis to see how to fit the cabin crew into that mold. (CC3)

How Is Competency-Based Cabin Training Conducted?

Based on the acceptance and wide use of AQP by airlines in the United States, the Federal Aviation Administration (FAA) gives airlines an exception so that they don't have to do standardized generic training. Airlines are allowed to customize their training content to address identified needs:

> What I have seen so far, in the US, almost all the major carriers do competency-based training. Some do it only for recurrent [training]. In the US it's very accepted ... one of the benefits for crew, as far as being more realistic, is if the airline would usually do one- or two-day recurrent [training] annually (before they would do one day with PowerPoints and you just want to drop dead) but now they will replace it and do two scenarios: one in the morning and one in the afternoon where the crews don't know what's going on. They're put in a simulator (like every day in real life) and maybe there is a fire or heart attack, they transition from doing theory (which some parts they can cover beforehand through e-learning) into making training very practical. They will do scenarios that will play out and instructors will play crew or an unruly passenger, then they'll do a debriefing. It makes the training go from very dry and theoretical to very hands-on. For the crew they like it a lot more because it's much more interesting than sitting around in a classroom. (CC3)

In other regions around the world, teaching practices that are called "competency-based" are being used. However, as they are not involved in a formal process, such as AQP, that tailors instruction to the unique challenges within the specific organization, "I would *not* say it's true competency-based—but a lot of airlines are moving into scenario-based training. So, even though they don't have the full competency-based approach, they can migrate a lot of boring classroom stuff into a scenario in the simulator (and they have cabin motion machines that look very realistic)" (CC3). In this way, through scenario-based practices, many airlines around the world are taking steps toward CBE.

Advantages and Challenges of Competency-Based Education:
Anecdotes from Industry

For the airlines that have begun incorporating CBE, an advantage is that it seems to reduce training times. This is made possible by the use of simulator scenarios that cover multiple training elements simultaneously. For example, a fire occurs and a passenger faints simultaneously, which requires trainees to demonstrate both first aid and firefighting skills (CC3).

However, there are also challenges associated with CBE. This approach is regarded as more labor intensive for airlines. Compared with classroom instruction, which a single instructor can easily facilitate, in a simulator several instructors are required. CBE requires more instructors and more time to prepare materials:

One challenge is manpower ... the cabin [simulator] isn't as big as a whole plane, but it's a chunk of cabin with a galley and lavatory, so you'll need different instructors in different places and there could be four, five or six trainees at the same time. You then need more instructors to complete the one-on-one observations. (CC3)

In addition, on the regulatory side, CBE is more labor intensive for inspectors. As one expert explained, "They have to review more data and it's more of a hands-on approach. Before they would just get a syllabus to sign off on, now they have to be there and see the training is being done to their satisfaction, as they are allowing them to do something outside of the usual regulations" (CC3). However, the same professional commented:

I think competency-based training will evolve. Scenario-based instruction is the first step that gets everyone in the mindset of thinking about how people are performing instead of just ticking the boxes. You go from thinking "I'll just teach two hours of this" to a scenario that might take one, two or three hours, but I am looking at the competencies. For airlines, it may seem scary or intimidating to put on a full competency-based training program. But, if you at least start with scenario-based training, you are on the way. The challenge for a lot of airlines is that the regulator has to be of a flexible mindset. In some cultures the regulator still has a very authoritative role. In that sense, it works easier in some countries. In some areas the regulator will say "I told you to do two hours, so do two hours." (CC3)

Conclusion

The perception of cabin crew has evolved through the years. Although many of these professionals still struggle with the balance between safety and customer service, recognition and respect for their work is increasing. Evidence of this is EASA's introduction of an "attestation" as a type of license for cabin crew. The attestation system has led to reduced training costs and increased professional mobility between European airlines.

With competency-based education (CBE) becoming increasingly common, either through complex AQP-type or more straightforward scenario-based curricula, many airlines around the world are moving toward this teaching philosophy. However, several challenges are associated with CBE, including the need for more instructional staff, an understanding of how to collect data through an established process and a supportive regulator.

It should be noted that acceptance and usage of CBE varies in regions around the world. "Many airlines are successful with this training approach and many others are well on their way" (CC2). However, in general terms, cabin crew training experts summarize the current state of CBE as follows:

As for competency-based [education]—how can I say this candidly?—I don't know if people have fully bought into the concept yet. They have been under the premise that time-based training was the preferred method because that was what the regulator was dictating. It's taking some time for that to change. (CC2)

Change will come from the regulators changing (because airlines can flip on a dime) and also the inspectors, because they are used to black and white and with competency-based [education] there is a lot of grey. For the inspectors that is difficult. Down the road I think it will be a bit of a domino effect and catch on. (CC3)

Chapter 6
Aircraft Maintenance and Competency-Based Education

> Experienced maintenance professionals still live in the world they grew up in. Most people don't understand what is meant by competency-based [education]. In some ways, it's like the Emperor's New Clothes; we always had it—it's a new word for an old thing.
>
> Interview, ENG4[1]

Introduction

A complex integration of personnel, equipment and information is required to maintain aviation operations. From an aircraft's regularly scheduled service, to a pilot-identified defect noted in the aircraft's flight log, to an aircraft that has been damaged in flight (e.g., heavy landing), aircraft maintenance personnel are the professionals responsible for returning the aircraft back to flight status. As there are regional differences in the terminology used to describe maintenance personnel, for simplicity's sake this chapter will use the term "engineer" or the acronym AMMTE (aircraft mechanics, technicians and engineers) to refer to all maintenance professionals.

A century ago, any individual could perform basic maintenance on an aircraft. Yet, modern-day aviation has grown so complex that it may not be feasible for one individual, or even one team of AMMTEs, to complete a job task. Teams—which are typically made up of licensed aircraft maintenance engineers (LAME), aircraft maintenance engineers (AME) and apprentices[2] at differing stages of their training—are responsible for initiating, removing, testing, replacing and certifying[3] aircraft components in accordance with documentation. In cases where these functions cannot be conducted within the team of engineers' roster, engineers must coordinate with and successfully hand over tasks to the next team of engineers.

---

1 ENG4: Maintenance Training Expert, Europe.

2 Please note that different regions around the world may use different terminology to describe these professionals.

3 Certification is restricted to appropriately licensed engineers who are rated on a particular aircraft.

Maintenance Engineers: A Brief History

If you think about it, aircraft engineering was the first aviation profession. The *Wright Flyer* had to be designed, built and tested prior to Orville Wright becoming the first pilot to successfully fly a heavier-than-air aircraft. One recalls a classic photo at the Virginia Air and Space Center of the Wright Brothers' wind tunnel, designed to assist the Wright Brothers' understanding of flight controls. This is testimony to the fact that engineering design was at the forefront of modern aviation. The development of a modern aircraft requires decades of work in engineering design prior to the first test flight. Getting to this point requires a sophisticated team, which for modern aircraft manufacturers can be a multinational operation.

Aircraft became increasingly sophisticated up to and beyond World War II. However, some of the greatest advances in engineering occurred with the advent of the jet aircraft. Jet aircraft were capable of flying at higher levels—generally above 25,000 feet (below that being the realm of turboprop aircraft)—and requiring far greater sophistication and integration of mechanical systems. Jet aircraft were capable of conducting transoceanic flights, which resulted in larger fuel capacity requirements. These larger fuel capacities required complex fuel transfer systems to maintain the balance of an aircraft in flight. As aircraft became larger, flight controls could no longer be operated with a simple cable system, and hydraulic systems were added to assist with the pilot's ability to effectively control the aircraft. In addition, for longer transoceanic flights, aircraft required sophisticated navigation systems. This led engineers to design, fit and maintain navigation systems (e.g., inertial laser gyros, OMEGA and non-directional beacons). Another consideration associated with jet aircraft was that as they were capable of flying at increasingly higher altitudes, they had to function at colder temperatures. Aircraft were encountering not only low-speed stall constraints but also limitations due to the speed of sound. The jet aircraft became an extremely complicated system to design and maintain, and this led to an increased need for highly skilled, licensed engineers.

Maintenance Engineer Licensing and Procedures

Even though licensing of pilots had begun decades before, it was not until 1948 that licensing of engineers occurred. Today:

> [the] use of multiplex systems and the digital bus further reduced all the electrical, instrument and radio categories into one avionics category. Meanwhile, developments in airframe and engine technologies were also changing the aircraft maintenance deployment pattern. The phenomenal improvement in on-wing reliability of large fan engines with the consequential reduction of on-wing engine maintenance resulted in the amalgamation of the airframe and engine categories into one mechanical category. This two-category system is

currently in use in many countries and is aligned with the European Aviation Safety Agency (EASA) licensing system. The EASA system went further to remove trade boundaries in the Categories. The Cat A, Cat B1/B2 and Cat C licences are structured along deployment of technical personnel in maintaining modern commercial aircraft. (Khee 2009, 11–12)

Basically, what most countries have today is a system that enables engineers to certify particular maintenance work functions. For example, under the European Aviation Safety Agency (EASA), a category of license allows an engineer to certify work (including the work of LAMEs, AMEs and apprentices). For example:

- Cat-A license allows an engineer to certify simple tasks such as wheels, brakes and seats (or cabin defects)
- Cat-B1 license enables an engineer to certify engine, airframes and also electrical work
- Cat-B2 license permits aircraft engineers to certify electrical and avionics maintenance; this certification requires the LAME to be able to confirm that the work was carried out per the *Aircraft Maintenance Manual* (AMM).

Engineers must also complete a "type course" and hours of experience on particular aircraft types. Similar to pilots, they must be "type rated" on the particular large aircraft they work on. For example, an engineer type-rated on the Boeing 737 who wishes to "certify for maintenance" on a Boeing 747 must complete an additional course on the Boeing 747 and then conduct "logged hours" of experience before earning a type rating on the Boeing 747.

Although engineers must "certify" aspects of the job task, it is interesting to note that certification of a job task does not actually require a LAME to do the work. It does, however, require the engineer to be able to certify that the job has been conducted in accordance with the prescribed procedure. In some cases this can be done at the completion of the job. For example, after a hydraulic pump has been replaced, an engineer can check that is has been fitted. In other cases, however, the task cannot be certified after the job task is completed. For example, if a particular torque wrench (a tool that can accurately apply a precise torque to a fastener such as a nut or bolt) is required, the engineer cannot certify whether the required torque has been applied to a particular bolt after the job task has been completed. Rather, at the step when the torque wrench is required to be used, the engineer must be present and complete that step (or, more likely, observe another person do it correctly). In this last case, the entire job must be segmented into components so that the certifying LAME is present when particular tasks are carried out. Furthermore, in regard to important aircraft systems, such as flight controls, there is a requirement for a "duplicate inspection." Here *two* licensed engineers are required to certify the work.

In the following excerpt, an experienced licensed engineer outlines some of these experiences:

In the ideal world there's one job to be done, you're there all the time. But in the real world there's so many jobs to be done within a short time frame, and you may have the pleasure of having several people do it for you. Or, some tasks may involve two people instead of one, or three, or an engine change—a heap of people at the same time. So, depending on the nature of the task, you have to know when to be there looking at a particular step of that task that cannot be inspected thereafter. Or, it can be done wrong. If you're experienced enough, you say "don't do this step until I am there" or "call me when you come to this step and then we will do this together." (ENG3)[4]

The Daily Task of a Maintenance Engineer

To better understand the maintenance competency framework, it is helpful to review the engineer's role and how engineers manage particular tasks. For example:

It's pretty much standard from the day you start. When you join a crew, the leading hand will be given the work package. … That package will contain the different types of maintenance tasks they've got to carry out, and they'll be looking at where they've got to be. So it might be tow an aircraft from a hangar and take it out to the engine run bay and do an engine run. They may have to do just a straightforward check on the aircraft. So there's a whole range of things. But you're getting to that mindset that, regardless of whatever work you're going to get, the first thing you do is "okay, what have I got to do, what safety equipment do I need, how do I render the aircraft systems to be safe so I can work on it?" (ENG2)[5]

If this crew were involved in accepting and dispatching aircraft from the terminal, a situation may arise where a pilot reports a possible "bird strike" (a potentially damaging collision with a bird) during flight. The engineers would first look around the aircraft to identify any potential indication of an impact. If a strike had occurred, it would be written into the aircraft maintenance log either as a "pilot entry" or "engineer entry." If the aircraft had been landing at the time of the bird strike, all flight controls (including flaps) would be inspected. This process of "inspection" is not based on a single engineer's preferences, but by referring to documentation (such as the aircraft manufacturer's manual). If the engineer must place a hand or equipment onto any flight control or any moving component, there is a requirement that the aircraft is first "made safe." For some aircraft, this would require the hydraulic system to be de-powered and warning signs placed in the aircraft flight deck stating "MAINTENANCE IN PROGRESS." After completion of the inspection manual requirements, a LAME would certify the aircraft as safe

---

4   ENG3: Training Manager, Southeast Asia and Europe.
5   ENG2: Training Manager, Southeast Asia.

to fly. This process is very procedural, but it is also informed by trade rules that are instilled in all engineers. As one engineer mentioned:

> That's one of the things that we say to apprentices, that the whole time you're working, you've got to be listening. You should be listening to the aircraft. Because if suddenly noises within the aircraft change, there's a reason for it, and pretty quickly you learn that when hydraulic power goes on, you will know what that sound is and you suddenly go "hang on, no one's told me hydraulic power's on," "hydraulic power's on, get yourself out of there." Particularly if you're working in flight controls. So it's that awareness that you've got to have too, and you've got to communicate. You've got to be talking with each other the whole time. (ENG2)

A more recent attribute of the engineer's job is competence to inspect one's own work. For example:

> If you had to lock wire two bolts, you lock wire it, then take your hat and turn it around and say "now I have done that job and I have to inspect it." That requires a different set of competencies because you need underpinning knowledge (what size of wire do I need, how many turns do I have to do to twist the wire, etc.). You're looking at your work from a different point of view. Then you say, "I agree, that is well done and it is in accordance with whatever regulations or specs," and you put the signature behind that job and take over responsibility for that job. That is the new world. (ENG4)

So, competent engineers require far more than just hands-on skills. There is an array of non-technical skills, such as attitudes, situation awareness and safety awareness, as well as working effectively with colleagues and knowing what one's actions mean to others. As another engineer outlined, not all trainees possess these non-technical skills or understand the impact of their actions on others:

> [One particular trainee] managed to get through some of the practical training, nothing to do with his hand skills so much, but he has the peculiar predilection where he has got some lapse(s) in inhibition. And he can't seem to help himself; or rather he helps himself to all the switches in the cockpit … so he goes into the cockpit, all this work is going on and he starts playing with the switches. And people go like, "Whoa! What the heck's going on?" And they find out that this guy's in the airplane, with the switches. (ENG3)

> I tell a lot of the students this is a field of endeavor where there are not meant to be trade secrets. … *You're supposed to share information.* If there's something new you learn you talk to your friends about it, you tell them about it and vice versa. … The day you think you're going to be one step ahead of your colleague by keeping vital information to yourself, that may affect safety of an

aircraft, is the day you and your family are going to be on the plane that he is certifying. (ENG3)

Historical and Contemporary Training Programs

In the early years, aircraft maintenance—which was basically airframes and engines—was left primarily up to the owners of the aircraft. "Early aircraft maintenance training programs were focused on mechanical systems and the engines as the aircraft was essentially made up of a structure, control mechanism and the engines" (Khee 2009, 9). As aircraft became increasingly sophisticated, especially with the introduction of electrical components and flight instruments, increased levels of specialization were required, with specialization centering on (1) engines, (2) airframes, (3) electrical and (4) instruments (avionics). The training of the aircraft engineer became a:

> ... very specialised trade. It is peculiar because of its demand for hands-on skills as well as knowledge of very complex systems. As a result, the training of aircraft maintenance engineers adopted the apprenticeship system, similar to the marine industry. The essence of an apprenticeship system is to learn a trade skill from the master craftsman. (Khee 2009, 10)

Over the last sixty years, many countries and airlines have conducted in-house training and certification of their engineers. However, for countries around the globe with less established aviation industries, the capacity to train the number of apprentices needed may require outsourcing to large training institutions overseas (e.g., the *Aer Lingus* training center in Dublin, Ireland).

As an example of the training footprint of maintenance engineers, a four-year training program will be discussed. Please keep in mind that this is a single example and that one should expect differences in training approaches in various geographic regions and between types of operations (e.g., airline, military and general aviation).

An engineer's training begins with theory-based classroom instruction, where students will specialize in a particular area of engineering (e.g., B1 or B2). The duration of theory-based instruction varies depending upon the specialization:

> For a mechanical guy there's 1,200 hours of theory or underpinning training (which is quite high for apprenticeships). We estimated avionics is around 1,300 hours of theory training. ... At the time we collected all this data a basic plumber was around 800 hours. So our trade is quite heavy on the theory, which I suppose is not a surprise considering the complexity of the machines we're working on. (ENG1)[6]

---

6   ENG1: Training Manager, Southeast Asia.

After completing the theoretical training, the trainee begins an apprenticeship, which is approximately three years of technical and further education (TAFE; with practical application either in a workshop or on the aircraft), followed by a practical fourth year, at the end of which the trainees have completed their trade. One engineer discussed the TAFE experience:

> We went out to the workshops and we cycled around every three months into different workshops. So the first two years of your four-year apprenticeship you went basically to the support workshops. I'll never forget, after nine months of being in the classroom and the intense theory and practical classroom training, my first workshop area was the fiberglass shop. I did three months in there working with the people in that area doing the little fiberglass jobs I could do. And then I rotated around all the other different sections at the time. You had the wheel bay and component maintenance and all sorts of pipes and hoses. It was really good, a really good system. By the time you get into the third year of your apprenticeship you were then starting to head into the aircraft sections and working in heavy maintenance and also out on line maintenance. (ENG1)

In addition to the specific training footprint of apprentice engineers, a specific experience record is required. This experience is not just total hours, but experience on particular aircraft subsections (generally split into chapters of a manual), such as airframes, structures, hydraulics and engines. To be licensed by a regulatory authority, an engineer must have demonstrated, via a schedule of experience (SOE), appropriate experience on particular aircraft components:

> You would have to show that you had experience in all areas across a particular aircraft type applicable to your trade. You would get a LAME, who has that rating for that aircraft, he would sign to say that you had been involved and carried out the maintenance tasks as per the maintenance manual in each chapter. And it said you had to remove, install, do a functional test. You had to get so many hours up, on your SOE. So each chapter would state how many hours of experience you had to have in that area. (ENG2)

However, there are challenges with this approach, when some maintenance training professionals find it unreasonable for trainees to repeatedly demonstrate generic skills on different aircraft types:

> Here you have a real clash because the regulator says you have to do this physically. I would say, if I have assessed the person can connect an electrical connector, I know they can do it because I have seen it on the mock-up and assessed it, why do I have to do it on a type-rated course particularly on the Boeing 747? Nowadays, you can set up a simulation to say, "well, we can have a virtual aircraft where we can move it and get access to it by clicking on the panel and the panel opens" instead of unscrewing 50 screws. Now, no airline would

allow you to remove 50 screws for training purposes. But, the requirement to do it physically is such. There is the conflict area. This is where big resistance comes from the industry. (ENG4)

Some of the greatest changes that are occurring within maintenance training today are with the use of simulation. Similar to pilots, engineers now use e-learning, part-task trainers and full-fidelity simulation to practice specific training activities that were traditionally done on an aircraft. For example:

> One of the great benefits that I've seen with the computer era is aircraft system training. There's some computer-based training aids that are out there, particularly for the later model aircraft like the A380 and the A330, where you can get computer-based programs that allow you to simulate, switching the hydraulic pump on, and you can actually see the flow of the hydraulics through a schematic going to the component. And you can switch it off and you can see the flow stop and you can fail a pump and see what happens. And just the computer model that's available in some of these later aircraft training packages are far superior to the flicking through pages and pages of reading that I did. ... You know it's just, it's fantastic, that sort of technology nowadays, it provides a far better training experience than in the old days with me. (ENG1)

Similar to the training environments in other aviation professions, instructional methods and technology are changing professional practice. Please refer to Chapter 10 for a longer discussion on simulation and training technologies. However, the traditional regulatory structure is focused on hours and a "check-the-box" approach to training, where trainees must physically demonstrate generic skills on each aircraft type. A shift towards competency-based methods is desirable to increase the training credit for simulation and other teaching technologies and to define generic skills that, once mastered, would not have to be repeatedly demonstrated within subsequent aircraft type ratings.

Competency-Based Education for Maintenance Professionals: A Global Framework

Competency-based education (CBE) for maintenance professionals "came from the pilot side and they are ahead of the maintenance side. The goal was to get away from time-regulated training towards competency-regulated training" (ENG4). Therefore, CBE should focus on trainees achieving competence rather than on the completion of a prescribed number of hours. "It's about what you did during those hours that matters, rather than the hours themselves. We need to know that you can do the job" (PIL9).[7]

---

7   PIL9: Instructional Designer, North America.

In 2007, an IATA Training and Qualifications Initiative (ITQI) meeting was held to discuss a human resource shortage of AMMTEs. This led to a 2008 study from the "ITQI Engineering and Maintenance kick-off meeting," which identified that the International Civil Aviation Organization (ICAO) licensing and training standards were not current with new aircraft technologies and training methodologies. It was agreed that a competency-based approach would improve this issue and that competency-based methods should be accommodated in Annex 1 of the Convention (ICAO 2006). This led to a 2011 amendment of the *Procedures for Air Navigation Services – Training* (PANS-TRG) ICAO document that included competency-based training and assessment guidelines for AMMTEs (ICAO 2006).

Compared with pilot competencies, the identification of competencies for maintenance professionals presented several challenges. For example:

> The pilot side is very structured because no matter the aircraft you fly, or the company you work for, there is a sequence to every trip. You prepare for flight, start the engines, pushback, et cetera, until you shut off the engines. That is a linear workflow. In maintenance it is completely different. All maintenance work has to be done in accordance with documentation—maintenance manual, job cards, service bulletins, you name it. In paper form, the types of work could be several feet [of paper] wide. There are thousands of jobs like that. How could we get them into a [competence] structure? When we started, everyone had their own ideas about how to structure the whole business. (ENG4)

The goal of the competency-based process was to align maintenance training with predetermined standards, identified through a task analysis. Aircraft manufacturers issue standard practices manuals (SPMs) to describe methods, tools/equipment and standards for tasks. It was determined that, regardless of aircraft type, maintenance practice on modern aircraft requires a set of *generic* competencies. When reviewing maintenance competency frameworks within the PANS document, the importance of documentation is evident, as competencies directly reference specific source material:

> If you look into the matrix in the PANS document, there is one column that says "what documents have to be used." If you find in there the fundamental skills, "standard practices manual" (SPM), then it's a generic task. Only if you see the "maintenance manual" (MM) it's very type-specific. (ENG4)

One interviewee pointed out that it can be helpful to think of generic maintenance competencies as a:

> … toolbox of skills. When you get to a certain job, you have to think, "what tools do I have to get out?" Some tools you will always need—like "I have to locate it, identify the part, get access, etc." There is probably no job you will

apply *all* the tools, except maybe an engine change. In comparison with the pilot, where there is a linear structure to their work, the mechanic has to decide (based on the component they are working on and the specific maintenance task) what "tools"—competencies—will be used. (ENG4)

It is clear that maintenance professionals must critically analyze their task and choose the right "tools" (competencies) for the job. Working with a "toolbox" of generic competencies is important because:

... there is no way you can perform all of the jobs in the maintenance manual during training. So, we have to think about it. How generically can we train people so they can work on whatever aircraft or component and fulfill the required standards? That is what we tried to do. ... Why should I teach someone to change pumps on all the aircraft types? It's always the same thing in a different location. But, the practices you apply are always the same. That is the basic message. If you can break it down to these core competencies and a person is able to do this, to have the required underpinning knowledge, then through application of maintenance documentation you should be able to do the job. ... *That is the key point—[CBE focuses on developing] professionals who can find information, read it and process it into craftsmanship.* (ENG4)

This has led to the creation of three categories of competencies:

1. aircraft (and engine) systems maintenance
2. aircraft structure maintenance
3. component maintenance (ICAO 2006, 72).

Beyond standard practices, there are core competencies that apply throughout maintenance work (and must be taught and assessed). Core competencies are associated with:

- maintenance resource management and threat and error management in maintenance
- work health and safety
- adherence to industrial standards and regulatory and company procedures (ICAO 2006, 72).

The competency frameworks for the three domains of maintenance professionals are presented in Table 2.1 in Chapter 2, and are discussed in further detail in Chapter 7. The competency frameworks were developed using generic information from aircraft and engine maintenance manuals, structural repair manuals, component maintenance manuals and actions described in standard practices documents. It is not expected that a trainee will master all of the competencies, but just those for a specific function selected by the regulator or approved maintenance organization they are associated with (ICAO 2006). Please note that engine maintenance uses the competence framework for either airframe systems or component maintenance.

Advantages and Challenges of CBE for Maintenance Professionals

One of the advantages of CBE for engineers is an *increased use of training simulators*. As one engineer explained:

> We are very lucky in [our company] because we've got these simulators and we can test competence in a simulator so that we don't necessarily have to run live aircraft. And the benefit about a simulator is you can simulate faults as well and watch how people react which you can't really do on a live engine run. (ENG1)

Another potential benefit of CBE is that instruction becomes *more targeted* toward trainee and organizational needs and less focused on hours. However, this is not meant to suggest that hours of experience as an apprentice are inconsequential. Well-designed CBE should incorporate apprenticeship opportunities and *promote practical experience*, although this can be a challenge in a real-world environment:

> The industrial environment in the airport aprons, hangars and workshops have changed with increasing pressure on turnaround times. Resources embedded in the system for teaching and mentoring trainees for the future workforce were slowly being eroded. *The traditional assumptions of regulators that trainees acquire hands-on skills and relevant practical work experience on the aircraft during the On-the-Job-Training (OJT) period is now in jeopardy.* Control of work schedule and assignment of trainees for jobs essential for their training as required for obtaining their AME licences are not priority items of the foremen or superintendents in the work areas. (Khee 2009, 13)

> When I started as an apprentice, we were always fiddling around with the plane. Nowadays they just inspect, it became basically a white-collar job. Still you have to change the oil, for sure. But 60–70% of the work is brain work. This is particularly an issue with the young generation. They won't have that experience. (ENG4)

If practical training is reduced too much, training may devolve into an artificially simplistic "tick-the-box" process. This "checklist" approach limits capacity to sign off trainees. Although this is not the intention of CBE, it is a pitfall that can occur within certain applications of the approach. For example, within Australia's CBE system, apprentices have to remove and install an undercarriage in order to complete their training. However, on modern aircraft, that is only done every 12–15 years. If there were a large number of apprentices, they may not all be able to participate in that task and a "log jam" could occur:

So in some ways, it's a step forward in that we now have to get everybody exposure across all the systems and all the tasks in their apprenticeship. The limitation is it's restricting the number of people you can put through the training pathway. (ENG2)

So you've got to put a tick in every box. In some ways we didn't allow for a transferability of skill. That was one of the arguments we had because of the way it ended up being structured: a person could be signed off having removed and installed a range of hydraulic system components, a range of pneumatic system components, but if they haven't done fuel, they can't be signed off. And if they haven't done water and waste, they can't be signed off. And you sit there and go, "well, gee, if a guy can be signed off in, say, three out of those four, does he really need the fourth?" When you stop and think about it he has had to apply: occupational health and safety, use documentation and manuals, render the system safe. Once he's done all that all he's doing is, say, removing four bolts and taking the pump off, putting a new pump on and doing the four bolts up. (ENG2)

A focus on trainees demonstrating generic competencies may ameliorate this issue. However, regulatory authorities must be mindful of how competency-based regulations associated with type ratings are worded, in order to ensure a focus on transferable generic competencies. Instead of requiring the physical changing of several components on a specific type of aircraft, it should be considered what type-specific tasks/task elements have to be applied and what tasks/task elements are of a generic nature and can be left out of type rating training:

Once a requirement is written in that structure, you're obliged to teach it to someone—which is nonsense. You have to look at it and say, how is it attached? If you are competent, you can definitely change this damn thing and you don't need the type-specific training. (ENG4)

Another limitation of some CBE programs is that they may be *reductionist*. When something as complex as expert professional practice is converted to a textual statement—a competency—it may become too narrow. The resulting CBE can limit intuition, experience, reflection and higher-order competence that is required for expert performance. The importance of training that targets "higher levels" of competence is explained by a senior engineer:

The higher level of competence can never be achieved by just signing these things up. It involves how you train the people, the process of training them, how you get them to that level. You can feed a person information, but the person is not able to continually take that mass of information, that's one aspect. The next aspect is to take it in and integrate it with whatever he knows already in a way that is meaningful and practically useful to him. So it's presenting information but in a way that the person can fit it in to his schema (whatever is already there). And that is the high level. ... [The result is] they come away really happy and it's very hard to do that with aircraft engineers. Because by nature, if you look at an accountant who's penny pinching, because of his training an engineer is a person who's critical about every bloody thing. Everything he's finding fault. (ENG3)

Conclusion

Overall, maintenance engineering is a highly specialized profession that requires a combination of theory, knowledge of complex systems, capacity to accurately utilize documentation, hands-on abilities and social teamwork skills. Maintenance training has traditionally used an hours-based approach to train and regulate the development of a trainee's proficiency and it can be expected that there will be resistance to a shift toward a competency-based model:

> We have to do a lot of work—to shift from "I can do it all by heart" to "you have to do it in accordance with the documentation." They don't want to give it up because they are proud and they may see the value of their education crumbling. If you can show that within 18 months you can do the same thing they did in five years, that's a shock to some people and they say, "no, that's not possible." That explains why there is a lot of resistance. (ENG4)

CBE has gained popularity in recent years, based on the consideration that instruction and experience can be targeted toward the generic and type-specific competence of the individual, rather than focused on hours completed. CBE also allows organizations to take advantage of modern teaching technologies, such as simulators. However, feedback from trainers with CBE experience indicates that there are potential pitfalls with this approach (such as training potentially becoming artificially simplistic and adopting the same "tick-the-box" checklist approach as traditional methods). Additional experience is needed on a global scale to guide the training industry to avoid these pitfalls and to define best CBE practices for aviation maintenance professionals.

Part II

References

Ball, A.L. 2013. History of the stewardess. *Travel + Leisure*, January.

Barry, K. 2007. *Femininity in Flight. A History of Flight Attendants.* London: Duke University Press.

Benjamin, R. 2013. Secretary General of ICAO address to the Regional Seminar in Preparation of ATConf/6. Hong Kong, January 28.

Breihan, J.R., A.E. Briddon, E.A. Champie, N.A. Komons, T.L. Kraus, P.A. Marraine, et al. 1998. *FAA Historical Chronology: Civil Aviation and the Federal Government 1926–1996.* Washington, DC: Federal Aviation Administration.

Dekker, S.W.A. 2003. Illusions of explanation: A critical essay on error classification. *International Journal of Aviation Psychology* 13: 95–106.

Dow, J. and H. Defalque. 2013. *The MPL – A Systems Approach to Pilot Training – Gains Acceptance by States and Industry.* Montreal, QC: ICAO.

EASA. 2012. Annex V: Qualification of cabin crew involved in commercial air transport operations. http://www.trafi.fi/filebank/a/1324448246/fc7d01828517 ce1f379484d995aa3294/4632-part-ara-ora-cc_-luonnos.pdf. Accessed March 25, 2015.

Elamiri, M. 2013. Plotting a path towards sustainable air transportation. *ICAO Training Report* 3(1): 3.

Eurocontrol. 2004. *European Manual of Personnel Licensing – Air Traffic Controllers* [Report No. 070504-01]. Brussels: Eurocontrol.

Eurocontrol. 2005. *Guidelines for Competence Assessment* [Report No. 050322-01]. Brussels: Eurocontrol.

Eurocontrol. 2015. *Specification for ATCO Common Core Content Initial Training, Edition 2.0.* Brussels: Eurocontrol.

Farrow, D.R. 2010. A regulatory perspective II. In *Crew Resource Management*, eds B.G. Kanki, R.L. Helmreich and J. Anca, 361–78. Amsterdam: Elsevier.

Flin, R., L. Martin, K.-M. Goeters, H.-J. Hormann, R. Amalberti, C. Valot, et al. 2003. Development of the NOTECHS system for assessing pilots' CRM skills. *Human Factors and Aerospace Safety* 3(2): 95–117.

Handy, B. 2002. Glamour with altitude. *Vanity Fair*, October. http://www.vanityfair.com/news/2002/10/stewardesses-golden-era. Accessed November 11, 2015.

Helmreich, R.L. and H.C. Foushee. 2010. Why CRM? Empirical and theoretical bases of human factors training. In *Crew Resource Management*, eds B. Kanki, R. Helmreich and J. Anca, 3–58. Amsterdam: Elsevier.

Helmreich, R.L., J.R. Klinect and J.A. Wilhelm. 1999. Models of threat, error and CRM in flight operations. In *Proceedings of the Tenth International Symposium on Aviation Psychology*, ed. R.S. Jensen, 677–82. Columbus: Ohio State University.

International Civil Aviation Organization. 2006. *Doc 9868 Procedures for Air Navigation Services: Training*. Montreal, QC: ICAO.

International Civil Aviation Organization. 2013a. *Doc 9995 Manual of Evidence-based Training*. Montreal, QC: ICAO.

International Civil Aviation Organization. 2013b. *International Pilot Training Consortium (IPTC) Pilot Core Competencies: A Way Forward*. Montreal, QC: ICAO.

International Civil Aviation Organization. 2014. *Cabin Crew Safety Training Manual*. Montreal, QC: ICAO.

International Civil Aviation Organization. (n.d.). *Air Navigation Commission: Approval of Amendment 4 to the Procedures for Air Navigation Services – Training (PANS-TRG, Doc 9868)*. Montreal, QC: ICAO.

Khee, L.Y. 2009. Evolution of aircraft maintenance training. *Journal of Aviation Management* 2009: 9–16. http://www.saa.com.sg/saaWeb2011/export/sites/saa/en/Publication/downloads/SAA_Journal_2009.pdf. Accessed May 4, 2015.

Klinect, J.R., J.A. Wilhelm and R.L. Helmreich. 1999. Threat and error management: Data from line operations safety audits. In *Proceedings of the Tenth International Symposium on Aviation Psychology*, ed. R.S. Jensen, 683–8. Columbus: Ohio State University.

Leung, W.-C. 2002. Competency-based medical training: Review. *British Medical Journal* 325: 693–5.

Lim, Y.K. 2009. Evolution of aircraft maintenance training. *Journal of Aviation Management* 2009: 9–16. http://www.saa.com.sg/saaWeb2011/export/sites/saa/en/Publication/downloads/SAA_Journal_2009.pdf. Accessed May 4, 2015.

Longridge, T.M. 1997. Overview of the advanced qualification program. In *Proceedings of the Human Factors and Ergonomics Society 41st Annual Meeting*, 898–901. Santa Monica, CA: Human Factors and Ergonomics Society.

MacKay, W.E. 1999. Is paper safer? The role of paper flight strips in air traffic control. *ACM Transactions on Computer-Human Interaction (TOCHI)* 6(4): 311–40.

Mavin, T.J. and P.S. Murray. 2010. The development of airline pilot skills through simulated practice. In *Learning through Practice*, ed. S. Billett, 268–86. Dortrecht, The Netherlands: Springer.

Mayerowitz, S. and D. Koenig. 2015. Rapidly-growing Asian airlines race to find qualified pilots. *New York Times*, February 4. http://www.nytimes.com/aponline/2015/02/04/us/ap-us-taiwan-plane-crash-pilot-shortage.html?_r=0. Accessed November 11, 2015.

Moroney, W.F. and B.W. Moroney. 1999. Flight simulation. In *Handbook of Aviation Human Factors*, eds D.J. Garland, J.A. Wise and V.D. Hopkin, 335–88. London: Erlbaum.

Nielsen, G.P. 1982. *From Sky Girl to Flight Attendant: Women and the Making of a Union.* Ithaca, NY: ILR Press.

Oprins, E., E. Burggraaf and H. Weerdenburg. 2006. Design of a competence-based assessment system for air traffic control training. *International Journal of Aviation Psychology* 16(3): 297–320.

Oprins, E. and M. Schuver. 2003. Competentiegericht opleiden en beoordelen bij LVNL [Competence-based training and assessment at LVNL]. *HUFAG Nieuwsbrief* 6: 2–4.

Reason, J. 1990. *Human Error.* Cambridge, England: Cambridge University Press.

Reason, J. 1997. *Managing the Risks of Organizational Accidents.* Aldershot, England: Ashgate.

Talbot, M. 2004. Monkey see, monkey do: A critique of the competency model in graduate medical education. *Medical Education* 38: 587–92.

Thomas, M.J. 2004. Predictors of threat and error management: Identification of core nontechnical skills and implications for training systems design. *International Journal of Aviation Psychology* 14(2): 207–31.

Todd, M. 2005. Cabin crew flying after victory over Virgin Blue. *The Age*, October 11. http://www.theage.com.au/news/national/cabin-crew-flying-after-victory-over-virgin-blue/2005/10/10/1128796469449.html. Accessed November 11, 2015.

Weitzel, T.R. and H.R. Lehrer. 1992. A turning point in aviation training: The AQP mandates crew resource management and line operational simulations. *Journal of Aviation/Aerospace Education & Research* 3(1): 14–20.

Whitelegg, D. 2007. *Working the Skies: The Fast-paced, Disorienting World of the Flight Attendant.* New York: NYU Press. E-book.

PART III
Practice and Implementation of Competency-Based Education in Aviation

Introduction to Part III
Practice and Implementation of Competency-Based Education in Aviation

Part III of this book moves away from theory into the practice of competency-based education (CBE). We have defined CBE as instructional design, training and assessment that systematically references competency texts. The following chapters describe how competency texts ("competencies") are written and explain instructional design and training practices for CBE as well as assessment techniques. The chapters of Part III also include brief introductions to the ICAO competency frameworks for design, training and assessment.

A major theme in Part III is the design and development of training, a task traditionally undertaken by instructional designers. One challenge in the development of CBE is that it creates a new role for instructional system design (ISD). ISD is based on a sophisticated body of knowledge that traditionally addresses:

- the analysis of jobs, tasks and learners to determine training needs
- analysis of the goals of learning relevant to those needs
- identification of training strategies aligned to these analyses.

However, the development of competency texts is also based on an analysis of jobs, tasks and functions. Therefore, the key elements of ISD may have already been undertaken in the competency-writing process. Does this mean that in a competency-based system the learning designer (ISD expert) is not required? We suggest that learning designers are still required within a competency-based system, but the role itself is different under a CBE approach than in a traditional ISD setting.

Within competency-based education, instructional designers may contribute to the specific tasks at the following levels of the industry:

- regulator level:
    - assist in the writing of competencies through analyzing jobs, tasks and functions and representing them in written statements
- organization level (instructional design tasks):
    - conduct fine-grained analysis of competencies written by regulators, perhaps to conduct a more sophisticated analysis to identify cognitive processes involved in performing the described task
    - analyze learner needs within a specific organization (identification of prior learning and current competence), as competencies may describe expertise already possessed by some learners

- analyze available learning resources and technologies (such as simulators) and drafting appropriate learning goals that may fall outside the scope of mandated competencies
- organization level (instructional development tasks):
  - analyze, select or develop resources, training strategies and systems to ensure that learners achieve the outcomes described in competencies
  - take advantage of the potential for innovative self-paced learning and the creation of well-designed learning resources and systems.

Therefore, although the traditional role of an instructional designer may not apply to competency-based methods, the knowledge base of ISD is clearly an important component of CBE.

Part III: Chapter Descriptions

Chapter 7: Designing Competencies in Aviation: reflecting the complex relationship between learning design and CBE, Part III commences with a chapter on the design and development of competency texts. The chapter covers part of the traditional ISD knowledge base in its presentation of techniques of job, task and function analysis. The chapter also covers approaches to competency text creation that do not draw on the ISD knowledge base, such as the popular DACUM approach, which rejects or sidelines the need for technical approaches to analysis associated with ISD. The chapter concludes with a discussion of some examples of competency texts from the aviation industry.

Chapter 8: Training: the next chapter concerns training strategies, resources and e-learning relevant to CBE. Because the focus on outcomes is a fundamental characteristic of CBE, trainers and training organizations are afforded considerable latitude for innovative delivery. In practice, limitations on resources, time and development expertise can mean that delivery in competency-based systems may not differ greatly from traditional approaches to training. In this chapter, we emphasize strategies and resources that tap into the flexibility inherent in CBE for innovative delivery.

Chapter 9: Assessment: following on is a chapter on assessment in CBE. Assessment in a competency-based system is closely tied to the content of competency texts. Assessment is an extremely important part of CBE systems for high-risk occupations such as those of the aviation industry. Aviation industry stakeholders need to be assured that holders of industry roles are competent, and it is the assessment process that underpins this assurance. The chapter covers the conceptual basis of competency-based assessment, as well as principles for generation of quality evidence and for making consistent assessment judgments.

Chapter 7
Designing Competencies in Aviation

Questions Considered in this Chapter:

- What is the focus of analysis for competency design?
- How are data about competence gathered?
- How are data about competence analyzed?
- How is analysis of data about competence represented?
- How does the design of competencies relate to instructional design?

Written statements or "competencies" that represent competent work are central to competency-based education (CBE). These competencies are used for designing, facilitating and assessing learning. Occupations that use the CBE approach adapt and evolve their own ways of conceptualizing competent work, their own ways of analyzing it and their own ways of structuring competency texts. The aviation industry brings together different occupational groups from different jurisdictions—air traffic control, cabin crew, aircraft maintenance engineers and pilots from multiple States—and a number of different approaches to creating competency statements/texts are used across the industry. In some cases a number of different approaches are used in the same occupation. In this chapter we look at how competencies are developed and written. To help understand the diversity of competency-writing practices in aviation, we look at established approaches to competency text production, including the two approaches identified in Chapter 1 (the technical and consensus approaches), before examining actual examples of competencies from the aviation industry. These competencies are the written statements that program designers, trainers and assessors in the aviation industry work with when they use the CBT approach.

Chapter 1 presented a brief overview of the process of analyzing competent work. It was pointed out that an early technique—going back at least to the beginning of the 20th century—for analyzing work was *task analysis*. This approach was formally initiated by Frederick Taylor (1906), who needed a method to capture the competent work of factory workers so that managers could rationalize, quantify and control the labor of their subordinates. The basic idea of breaking work down into efficient chunks or tasks—the "division of labor"—had been introduced early in the industrial revolution. The practice and its potential payoff for factory owners were described by Adam Smith and others in the 1700s. But the idea of analyzing and representing work for the purpose of more efficient management is an innovation of the 20th century. Taylor and others promoted the observation-based methodology of "time and motion" studies to analyze tasks. This process involved painstaking, minute description of the activities of workers to determine optimal activity sequences and timeframes. It was hoped that managers could use this knowledge to eliminate inefficiencies and shift the locus of control of work from workers to managers.

A specialized discipline of task analysis has emerged since then, with its own traditions, knowledge base, experts, research and debates. The field of task analysis provides concepts to inform the job of writing competencies. For our purposes it will be helpful to distinguish four phases of task analysis in the process of creating competency texts: selecting a focus for analysis, gathering data about competent work, analyzing the data and representing competence in structured written statements/texts.

Focus of Analysis

For Taylor (1906), a task was simply a discrete activity performed by workers. A number of such tasks constituted a "job." It was assumed that an occupation could be reduced to a set of tasks that were somehow linked together. More sophisticated ways of understanding the focus of analysis have emerged in the past century. One development in the idea of task analysis is linked with the rise of systems approaches. For example, Kirwan and Ainsworth (1992) defined task analysis "as the study of what an operator (or team of operators) is required to do, in terms of actions and/or cognitive processes, to achieve a system goal" (1).

A different emphasis may be seen in cognitive task analysis (CTA). Crandall, Klein and Hoffman (2006) explained how the "task" is conceptualized in CTA:

> It may seem straightforward to think about "task" as people engaged in discrete activities or sequences of activities aimed at achieving some goal. This is a traditional notion of "task." But in complex cognitive systems, it is not always the literal action sequences—the steps—that matter as much as the fact that practitioners are trying to get things done; they are not simply performing sets of procedures. Therefore, we define tasks in this broader sense as the outcomes people are trying to achieve. (3)

Alternatives to a Focus on Tasks

Alternative focuses for analysis have been proposed. Researchers concerned with analysis of jobs for labor market, employment services, career counseling and training purposes have criticized the traditional task analysis approach for neglecting the way tasks are combined in actual work settings. An American alternative that responds to this criticism was developed by Sidney Fine and colleagues, with Fine and Wiley's seminal *Introduction to Functional Job Analysis* appearing in 1971. Fine's approach reimagined tasks as components in a "work-doing system" rather than a "job." Analysis of the work-doing system involved the analysis of three elementary components: the worker (e.g., capacities, experience), work organization (e.g., purpose, goals) and work (e.g., worker functions, worker instructions). For Fine and Cronshaw (1999), "task analysis" is the analysis of work, "organizational analysis" focuses on work organization and "qualifications analysis" is used to analyze workers. Functional job analysis involves all three forms of analysis, and the traditional task analysis focus is treated as a component within the overall analysis.

In Britain, Mansfield and Mathews (1985) created the "job competence model," which was explicitly oriented to CBE. Their model says analysis should focus on "work roles," which consist of four components:

- technical expectations
- managing contingencies
- managing different work activities
- managing the interface with the work environment (Mansfield and Mitchell 1996, 50).

In this model, the "technical expectations" component corresponds to the focus of traditional task analysis. But in the job competence model, three types of "managing" complement technical competence in a whole work role. A sufficient analysis should address all four components of competence.

Mansfield (1989) proposed a method of analysis—"functional analysis" (not directly related to Fine's model)—specifically to address a holistic understanding of work roles reflected in the job competence model. The aim of the functional analysis approach was to produce a statement of standards rather than descriptions of tasks. In making a case for functional analysis, Mansfield made a distinction between his approach and Fine's functional job analysis model. Mansfield explained that his own functional analysis approach possessed three essential features:

- the focus on whole work roles, which promotes a broad view of occupational competence
- an outcome approach, which identifies key purposes and functions, drawing together different activities, which are described in different ways (and which may be performed using different methods and techniques), but have the same purpose and the same standards attached to them
- a "top down" method, starting from a clear functional statement of the entire occupational area, breaking this down into significant roles (Mansfield 1989, 5).

While researchers with a background in employment services or career counseling have generally favored more complex understandings of competence, a simpler approach is advocated by proponents of the "DACUM" approach pioneered by Canadian social services specialists. Their approach (covered in more detail later) yields "job profile charts" that differentiate tasks and *duties* that make up a job.

Clearly, the idea of "task" as a focus for job analysis has undergone substantial evolution since Taylor (1906) first offered it as a way of understanding work. We have seen that the concept of task has been recast in terms of systems theory, remodeled in cognitive terms and reimagined in terms of functions and whole-work roles.

In the context of the aviation industry, a number of these conceptualizations of task analysis are in play. For instance, in some frameworks (e.g., the multi-crew pilot license [MPL] competencies), the focus of analysis is on phases of flight (a systems-oriented task approach). The evidence-based training (EBT) framework presents a more holistic model that produces "core competencies" that are found in a number of aviation roles, although they manifest differently in each role. The ICAO aircraft maintenance competencies, on the other hand, focus on functions.

How Data about Competence are Gathered

No matter what is taken to be the appropriate focus of analysis for competency text design, data must be collected to provide content for the written statements/texts. At this stage, a difference can be observed between "technical" and "consensus" approaches to competency text production. The technical approach aims for rigorous data collection, generally undertaken by specialists from outside the focus occupation. We will consider the methods of the technical approach first before looking at consensus models such as "DACUM" (a contraction of "design a curriculum"). Consensus models derive content through a panel of occupational experts, whose knowledge is taken to be sufficient for determining the content of competency texts.

Technical Approaches to Data Collection

In terms of rigorous techniques for data collection, two broad categories can be distinguished: observation and eliciting information from human subjects who know the task. Taylor (1906) promoted observation techniques as an objective way of determining the nature of competent work. Kirwan and Ainsworth (1992) described contemporary observation techniques such as "activity sampling," in which the goal is to determine how much time is devoted to each activity that is part of the task; and "direct observation," which seeks to "capture all the significant visual and audible events, in a form which makes subsequent analysis reasonably easy, and which does not influence the performance of the task" (54). The last part of the quote refers to a challenge created by the analytic process itself: if observation disrupts the work itself in any way, the work may be misrepresented. Therefore, technical approaches must be implemented with care to minimize disruption.

In contrast to observation-based techniques, subject-based approaches extract task data directly from workers (or "operators"). Subject-based techniques include the "critical incident" technique, questionnaires, structured interviews and "verbal protocols," which are all sophisticated methods administered by trained specialists. The critical incident technique, which was described by psychologist John Flanagan, had its origins in aviation. Flanagan (1954) explained that the method arose from reflecting on the process by which beginning pilots were eliminated from flight training schools in the United States Army Air Forces in the early 1940s. The reasons for failing these novice pilots were written down, and sometimes the information contained in these records was useful for the development of selection processes and training programs. Flanagan and others used information about failures to analyze pilot work. Flanagan's technique involved the "systematic effort to gather specific incidents of effective or ineffective behavior with respect to a designated activity" (328). For example, a study to design combat leadership training gathered information from combat veterans. The veterans were asked to:

> report incidents observed by them that involved behavior which was especially helpful or inadequate in accomplishing the assigned mission. The statement finished with the request, "Describe the officer's action. What did he do?" Several thousand incidents were collected in this way and analyzed to provide a relatively objective and factual definition of effective combat leadership. The resulting set of descriptive categories was called the "critical requirements" of combat leadership. (Flanagan 1954, 329)

Flanagan (1954) argued that task analysis should focus on the determination of such critical requirements. The critical incident technique continues to be promoted as an effective way of analyzing tasks. David McClelland (1991) developed a variation of this technique, which is called the behavioral event interview (BEI). BEI combines Flanagan's technique with the thematic apperception test, a psychological testing method that involves asking subjects to interpret pictures of people undertaking unspecified tasks. BEI asks interviewees to identify and discuss successes and failures they know of in their occupational specialty. The BEI technique allows the characteristics of competent people to be determined, rather than the "task elements of jobs" yielded by the traditional critical incident technique.

Other subject-based approaches to task data collection include questionnaires, interviews and verbal protocols. Questionnaires (surveys) involve a set of questions for competent operators. Operators may be invited to respond to multiple-choice questions, rating scales or open-ended questions about their work. Questionnaires can be distributed to a large number of operators, but because the approach is relatively rigid, only a limited amount can be learned about the task in this way. In contrast, interviews with subjects have scope to gather more nuanced information. Structured or unstructured interviews can be used to gain insight into competent work. The verbal protocol technique involves having operators "verbalize" as they perform their work, giving the task analyst a "live" picture of the work. Like interviewing, this technique has the potential to reveal nuanced data about the task.

The formal techniques of observation, activity sampling, critical incident technique, questionnaires, interviews and verbal protocols can be used individually or in different combinations for gathering data about the task(s) in question. These techniques have become highly sophisticated; they usually require skilled analysts to implement, and the task analysis entails substantial time and costs.

Consensus Approaches to Data Collection

An alternative approach to gathering task data, called the DACUM, emerged in Canada in the early 1970s. Canadian social services consultants wanted a method of task analysis that was more economical than that offered by the technical approach:

> The DACUM technique is an effective method of determining quickly, at relatively low cost, the tasks and duties expected of anyone employed in a given job or occupation. The premise is that people employed in various occupations are capable of describing the knowledge, skills, and attitudes required to work in them. (Joyner 1995, 248)

The DACUM technique involves a trained facilitator and a group of eight to twelve expert workers from the occupation in question, working together over two or three days. During the workshop, the experts reach consensus about components of their occupation. The DACUM process produces a "DACUM chart," which is a comprehensive list of the tasks and duties that are typically performed within the occupation. While the task outcome of the process has a traditional focus, the duty component concerns responsibilities that may relate to multiple tasks or extend beyond the scope of those tasks (Joyner 1995).

Observation and subject-based techniques require trained analysts to stand outside the occupation and look in, using rigorous, time-consuming and often costly methods to determine the requirements of jobs and the characteristics of competent operators. In contrast with this technical approach is the consensus approach of the DACUM technique. Consensus approaches are based on the assumption that operators can directly articulate the requirements of jobs and their own competence.

How Data about Competent Work are Analyzed

The results of observations, interviews or expert panel approaches such as the DACUM technique must be analyzed to produce information appropriate for use in competency statements/texts. Approaches to analysis range from the highly sophisticated (associated with technical approaches) through to the pragmatic (associated with consensus). The first step in analysis is often some form of arrangement (or mapping) of the data that has been gathered about jobs and tasks. Early techniques, such as those developed by Gilbreth and Gilbreth (cited in Kirwan and Ainsworth 1992), involved representing work in the form of "process charts." Kirwan and Ainsworth (1992) explained that "these charts were top-down flow diagrams of the task in which each behavioural element was classified and then represented by a particular symbol" (84). Gilbreth and Gilbreth proposed a set of 53 standard symbols for discrete types of behavior. In 1972 the American Society of Engineers reduced this to a set of five, which is still in use for preparing flow charts. Other charting techniques for presenting task data are input–output diagrams, functional flow diagrams, information flow charts, Murphy diagrams, critical path analysis, Petri nets and signal flow graphs (Kirwan and Ainsworth 1992).

Annett and Stanton (2000) explained that as analysts attempted to represent more complex tasks, alternative techniques such as hierarchical task analysis (HTA) evolved. This evolution coincided with a move away from behavioral perspectives on skilled work (which focus on observable stimuli and responses, such as the behaviors represented in a flow chart) to cognitive approaches (which included information about cognitive processes involved in competent work):

> The key feature of HTA is that *tasks*—those things that the person is seeking to achieve—are essentially defined by *goals* rather than actions as such, and that complex tasks may be analysed by decomposing a hierarchy of goals and sub-goals. (Annett and Stanton 2000, 2)

As this definition suggests, understanding the goals related to the tasks allows a more complex analysis, compared with analysis based on describing observable actions. Instead of analyzing tasks using linear representations of actions, HTA uses component and overarching goals, allowing simpler elements to be distinguished within complex task wholes.

The tasks may be analyzed further by using cognitive task analysis (CTA). This additional step focuses on identifying workers' *mental processes* related to the goals identified by the HTA process. According to Clark, Feldon, van Merriënboer, Yates and Early (2008):

> Analysts use CTA to capture accurate and complete descriptions of cognitive processes and decisions. The outcome is most often a description of the performance objectives, equipment, conceptual knowledge, procedural knowledge and performance standards used by experts as they perform a task. (1)

Although there are many variations on the CTA technique, five steps are common (Clark et al. 2008). The first corresponds with Kirwan and Ainsworth's (1992) task data collection stage, in which information about the task is collected through observation and/or some form of interview. The second step aligns with Kirwan and Ainsworth's task description techniques, which include charting, diagrams and the HTA technique.

Assumptions about differences between types of knowledge come into play in a complete CTA. Clark et al. (2008) distinguished between "declarative knowledge" and "procedural knowledge." They defined declarative knowledge as:

> hierarchically structured propositional, episodic, visuospatial information that is accessible in long-term memory and consciously observable in working memory. ... This type of knowledge supports performance through the conceptual understanding of processes and principles related to a task and the role that the task plays within its broader context.

According to Clark et al. (2008), competent operators may be consciously aware of declarative knowledge related to a task, and therefore may be able to articulate it directly so that the cognitive determinants of task performance can be formally recorded. In contrast, procedural knowledge is the knowledge of how to do a task. Procedural knowledge may not be accessible to the consciousness of the worker in possession of it, because "automaticity" has developed over time. That is, tasks that the operator may have initially pictured in terms of declarative knowledge may be transformed into an unconscious and "holistic" action. Therefore, the reports of experts cannot necessarily be relied on to produce complete exposés of procedural knowledge. The involvement of an appropriately trained analyst may be necessary to identify the full range of components of procedural knowledge.

In addition to the declarative and procedural knowledge necessary for task performance, CTA also identifies "cues" (contextual triggers for particular actions) and "decision points" (at what point in the task to act on cues; Clark 2008). Finally, cognitive and psychomotor skills triggered by cues at appropriate decision points must also be described, although automaticity tends to obstruct the analysis of these skills here as well.

Alternative ways of breaking down cognitive processes have been proposed by other education and training researchers. For instance, Gagné and colleagues (cited in Smith and Ragan 2005), proposed the following breakdown:

- declarative knowledge (being able to recall terms or definitions)
- discriminatory knowledge (knowing how to identify relevant differences)
- conceptual knowledge (knowing how to explain and apply concepts)
- procedural knowledge (knowing how to undertake procedures)
- problem-solving (knowing how to deal with non-routine situations)
- cognitive strategies (knowing and manipulating one's own cognitive processes)
- attitudinal knowledge (ability to adopt appropriate attitudes)
- psycho-motor skills (ability to "smoothly" perform physical processes).

Bloom's "taxonomy" presents another approach to breaking down the learning required for people to complete tasks. It is based on research that began to appear in the 1950s. Bloom et al. (1956) conceptualized the broad field of possible learning outcomes into three domains: cognitive (knowledge), affective (attitudes) and psycho-motor (skills). Within each of these domains a hierarchy of learning outcomes is proposed, from simple to more sophisticated. These hierarchies have evolved over a long period, with a major revision published in 2001. Table 7.1 presents an overview of levels of learning outcomes for each of the three domains conceptualized by Bloom and colleagues. Each of these outcomes aligns with learning strategies.

Table 7.1   **Extended and Updated Synthesis of Bloom's Taxonomy**

Complexity	Cognitive Domain (Anderson & Krathwohl, 2001)	Affective Domain (Krathwohl, Bloom & Masia, 1964)	Psychomotor Domain (Harrow, 1972)
Simple	Remember	Receiving /Attending	Reflex movements
	Understand	Responding	Basic-Fundamental Movements
	Apply	Valuing	Perceptual Abilities
	Evaluate	Organisation	Physical Abilities
	Create	[Internalisation of Value]	Skilled Movements
Complex			Non-Discursive Communication

The frameworks for analyzing learning tasks developed by Gagné and colleagues and by Bloom and colleagues offer relatively sophisticated ways to understand the intended level of complexity of a competency text, create appropriate learning tasks and allow decisions to be made about the most effective learning strategies. Although analyzing a learning task described in a competency text may be easier using the simple knowledge, skills, attitudes (KSA) framework, the more fine-grained alternatives have the advantage of allowing greater fine-tuning of training strategies. A comparison of the KSA and alternative frameworks is shown in Table 7.2. Note that the frameworks of Gagné et al. and Bloom et al. demand more complex decisions about the learning outcomes implicit in the learning task, although the pay-offs are worthwhile.

Table 7.2   Comparison of Frameworks for Analyzing Learning Tasks (note that bracketed sets of learning types increase in complexity from top to bottom)

Traditional model	Gagné and colleagues' model	Bloom and colleagues' model
Knowledge	Declarative Knowledge Discriminative Conceptual Knowledge — *Intellectual Skills* Principles	Remember Understand Apply Evaluate Create — *Cognitive Domain*
Skills	Procedures Problem-solving Cognitive strategies Psycho-motor skills	Reflex movements Basic movements Perceptual abilities Physical abilities Skilled movements Non-discursive communication — *Psychomotor Domain*
Attitudes	Attitudes	Receiving /Attending Responding Valuing Organisation Internalisation of Value — *Affective Domain*

The DACUM technique may adopt a different approach to analysis. The original DACUM method distinguished between collecting data about tasks and duties (through expert consensus achieved during the DACUM workshop) and a subsequent phase of sophisticated analysis. Joyner (1995) explained that while no trainers or educators are supposed to be present during the workshop (only a trained DACUM facilitator), educators *are* consulted afterwards to analyze the DACUM chart and determine the underlying "knowledge, learning, skills, and attitudes required to attain mastery" (248). Potentially, a CTA can be performed on the data collected during the workshop. But Norton (2004) promoted an influential variation in which all underpinning KSAs were identified during the DACUM workshop itself:

> While the knowledge, skills, tools, and worker behaviors are not tasks, they are enablers which make it possible for the worker to be successful. Because these four enablers are so important, considerable attention is given during the DACUM workshop to identifying lists of each. Because these attributes are different and distinct from the tasks, it is very important to keep them separate if a high quality analysis of job performance is to be obtained. (Norton 2004, 4)

Norton's (2004) version of the DACUM approach assumes that experts have access to all significant knowledge as well as sufficient insights to analyze a task. However, research reviewed by Clark et al. (2008) showed that experts can overlook important steps of tasks, indicating that the consensus approach to task analysis may produce incomplete representations of competence. Presumably, skilled DACUM facilitators may overcome this possible limitation.

Just as different ways of conceptualizing the focus of analysis are evident across the aviation industry, there are different approaches to data collection and analysis. The *Cabin Crew Safety Training Manual* (ICAO 2014) explains that the competency framework presented in the manual:

> was developed by the ICAO Cabin Safety Group, composed of subject matter experts from States, airlines, manufacturers and international organizations, through a process of consensus on what constitutes cabin crew competencies, necessary for safe operations. The content of the framework represents the result of this exercise and is an internationally agreed upon baseline for cabin crew competencies. (ICAO 2014, 3-2)

This approach was explained in some detail by a professional interviewed for this book (PIL9).[1] The interviewee explained that when a new ICAO competency framework is to be developed, the ICAO will:

- *Assemble a group of experts.* These experts include regulators, industry representatives and training providers. In some cases, these experts are also expert practitioners in their domain.
- *Hold a series of meetings.* The objective of the meetings is to develop a competency framework that conforms to the established format (competency unit, competency element, performance criteria, and evidence and assessment guide if applicable).
- *Use a systematic methodology.* The experts are instructed to apply three basic principles when developing competency frameworks:
  1. To remove their "instructor hat" and to describe what they see as exemplary job performance. In other words, describe what you want to see on the job and not what you teach in the class.
  2. When describing job performance, describe with an action verb. The action verb is used to focus on an *observable* action or outcome.
  3. Consider what is generic in the description. In other words, whatever is described as job performance should be understood and applicable across the world. (PIL9)

The interviewee added that:

> The development of competency frameworks involves an iterative process over a series of two- to three-day meetings. In some cases, it can take much longer. For example, the aircraft maintenance framework took over two years to complete, with several face-to-face and virtual meetings. (PIL9)

---

1   PIL9: Instructional Designer, North America.

This consensus approach to determining competencies reflects principles of the DACUM method advocated by Norton (2004). A different approach was used for aircraft maintenance competencies. To derive these competencies, "a functional/ task analysis was carried out for several generic maintenance functions" (ICAO 2006, 4-A-1). Once these functions were determined, a literature search was used to detail the competencies. The competency frameworks developed through this process:

> list the competencies for three domains: aircraft systems maintenance, aircraft structures maintenance and aircraft components maintenance. The frameworks were developed by combining the existing generic information found in aircraft and engine maintenance manuals, structural repair manuals, component maintenance manuals and the actions described in standard practices documentation. (ICAO 2006, 4-2-1)

Representing Competence in Written Statements/Texts

Although competency texts are common to all implementations of CBE, there is substantial variation concerning the most appropriate way to represent competence. For example, competency texts vary in terms of:

- *Amount of detail.* Looking across different occupations that have adopted the CBE approach, one obvious difference is the amount of detail contained in the texts. Some have a simple structure and/or a relatively brief overview of the competency. Others can contain 20 or more different categories of information about the competency, providing a highly prescriptive description.
- *Occupational focus.* The texts may emphasize either behavioral information or information about cognition and/or the social practice of the occupation. Many formats include more than one kind of information.
- *Scope.* The scope of the texts varies. Some competencies refer predominantly to the competence itself, while others contain additional information such as guidance for training and assessment.

When we look at the aviation industry with its range of component occupations, it is possible to identify variations on each of these dimensions. However, one format in particular is promoted by the ICAO in most of its competency frameworks to date. This format is heavily influenced by behavioral objectives theory, which was briefly considered in Chapter 1. An early advocate of behavioral objectives was Robert Mager, who explained the rationale as follows:

> Simply put, a usefully stated objective is one that succeeds in communicating an intended instructional result to the reader. It is useful to the extent that it conveys

to others a picture of what a successful learner will be able to do that is identical to the picture the objective writer had in mind. ... What you are searching for is that group of words or symbols that will communicate your intent exactly as YOU understand it. (Mager 1962, 19)

According to Mager, a "usefully stated objective" will include three kinds of information:

1. *Performance.* An objective always says what a learner is expected to do; the objective sometimes describes the product or result of the doing.
2. *Conditions.* An objective always describes the important conditions (if any) under which the performance is to occur.
3. *Criterion.* Wherever possible, an objective describes the criterion of acceptable performance by describing how well the learner must perform in order to be considered acceptable. (Mager 1962, 21)

The components of a behavioral objective, as identified by Mager (1962) and others, form the basis of a number of large-scale implementations of CBE, including the British National Vocational Qualifications (NVQ) system in the 1980s, the Australian National Training System (NTS) in the 1990s and now the ICAO. In these contexts, the practice has been to record task components in terms of "elements" and criteria in terms of "performance criteria" tied to individual elements, and to indicate conditions at some point in the documents. In the ICAO and other models, the performance and criterion categories identified by Mager predominate. However, models like the ICAO and Australian NTS implementation of CBE may also contain information about cognitive processes relating to the behaviors described in elements and performance criteria.

Cabin Crew Competencies

The ICAO model can be found in frameworks developed for a range of occupations in the aviation industry. For example, in 2014, the ICAO released its *Cabin Crew Safety Training Manual*, which outlines a competency framework for cabin crew training. Each sub-framework focuses on a different safety-related aspect of cabin crew work:

a. normal operations
b. abnormal and emergency situations
c. dangerous goods
d. cabin health and first aid
e. security threat situations (ICAO 2014, 3–4).

Each of these areas includes a number of competencies. For instance, "normal operations" is articulated in the following competency units:

1. Perform duties and responsibilities during ground and pre-flight operations.
2. Perform duties and responsibilities during pushback and taxi.
3. Perform duties and responsibilities during take-off.
4. Perform duties and responsibilities during climb.
5. Perform duties and responsibilities during cruise.
6. Perform duties and responsibilities during descent and approach.
7. Perform duties and responsibilities during landing.
8. Perform duties and responsibilities during post-landing and post-flight operations.

Each of the competencies within the different cabin crew safety frameworks uses the elements and performance criteria format, with additional categories of information included. For example, "Competency Unit 1.: Perform duties and responsibilities during ground and pre-flight operations" contains the following elements:

1.1. Perform planning duties.
1.2. Participate in flight crew and cabin crew briefings.
1.3. Perform preflight checks.
1.4. Perform passenger boarding and pre-pushback duties.
1.5. Manage abnormal or emergency situations.
1.6. Communicate with flight crew, other cabin crew and passengers.

For "Competency element 1.1.: Perform planning duties," the performance criteria are:

1.1.1. Report for duty.
1.1.2. Obtain applicable information/documentation.
1.1.3. Review documents required for the flight.
1.1.4. Update documents required for the flight, if applicable.
1.1.5. Check minimum cabin crew complement.

In a departure from the behavioral objectives approach, the manual also identifies cognitive components that underpin the performances described in the elements and performance criteria. For example, the following knowledge and skills are specified for "Competency Unit 1.: Perform duties and responsibilities during ground and pre-flight operations":

Knowledge:
a. system/method used to report for duty
b. regulatory requirements regarding specific items required for duty
c. types of documents and information required, where/how to obtain them, and how to complete and/or update them
d. minimum cabin crew complement for each aircraft type, in accordance with the applicable regulations.

Skills:
   a. workload and time management
   b. planning and coordinating resources (for in-charge cabin crew member).

The ICAO competencies for cabin crew therefore represent a hybrid model that covers behavioral and cognitive components of competence. In preparing these competencies for cabin crew, the ICAO did not consider the aspects of the work concerning comportment toward passengers, although communication with passengers and other crew is addressed at several points in the framework (see Chapter 5 for additional information on cabin crew competencies).

Aircraft Maintenance Competencies

Competencies for aircraft maintenance personnel have been published by the ICAO in the 2011 *Amendment No. 1 to the Procedures for Air Navigation Services* (ICAO 2006). For systems maintenance personnel, the competencies are:

1. Perform fault isolation.
2. Perform maintenance practices.
3. Perform service.
4. Remove component/assembly.
5. Install component/assembly.
6. Adjust.
7. Test.
8. Inspect.
9. Check.
10. Clean.
11. Paint.
12. Repair.
13. Perform MEL and CDL/DDPG procedures (minimum equipment list [MEL]/configuration deviation list [CDL]/dispatch deviation procedures guide [DDPG]).

For aircraft structure maintenance personnel, the competencies are:

1. Perform aircraft structural repair inspection.
2. Perform structural damage investigation, cleanup and aerodynamic smoothness check.
3. Perform special process application.
4. Perform metal rework/testing.
5. Perform structural repair.

For aircraft component maintenance personnel, the competencies are:

1. Perform testing fault isolation.
2. Perform disassembly.
3. Clean.
4. Perform inspection/check.
5. Repair.
6. Perform assembly.
7. Perform storage (transportation).

Some of these competencies can be broken down into elements. For example, the elements of the aircraft components maintenance competency "1. Perform fault isolation" are:

1.0. Recognize and manage potential threats and errors.
1.1. Prepare for fault isolation—collect fault data.
1.2. Verify fault data.
1.3. Develop fault isolation procedure.
1.4. Perform fault isolation procedure.
1.5. Define fault rectification procedure.
1.6. Complete fault isolation.

These elements, in turn, have specific performance criteria. The performance criteria for "1.1. Prepare for fault isolation—collect fault data" are:

1.1.1. Collect fault data from relevant aircraft technical logs (printed or electronic) of pilot or maintenance reports—if available.
1.1.2. Collect data from aircraft recorders/in-flight transmitted records (maintenance messages).
1.1.3. Collect fault data from maintenance defect reporting sheet.

Knowledge and skill statements are not provided for the aircraft maintenance competencies, making these competencies more behavioral than the cabin crew competencies considered above. Please refer to Chapter 6 of this book for additional information on competencies within this profession.

Pilot Competencies

Work done by the FCLTP on developing competencies for entry-level training and assessment for pilots produced nine "competency units" (please refer to Chapter 4 of this book for more details). The analysis for these competencies took "phases of flight" as the guiding principle for determining tasks for analysis. These nine competencies are:

1. Apply threat and error management principles.
2. Perform ground and pre-flight operation.
3. Perform take-off.
4. Perform climb.
5. Perform cruise.
6. Perform descent.
7. Perform approach.
8. Perform landing.
9. Perform after-landing and post-flight operation (ICAO 2006, Amendment No. 1, 2-2).

The exception to the approach used to determine tasks based on phases of flight is the first competency on the list, "Apply threat and error management principles." As noted in Chapter 4, the principles of threat and error management are important knowledge for pilots because risks can never be completely eliminated from aviation. The proactive option was to manage risk at multiple levels in the industry. Competencies such as "Apply threat and error management principles" are often referred to as "generic" or "transversal" competencies—that is, skills and knowledge that can be applied in a range of situations (Trzmiel, Hee Kim, Brennan Kemmis and Becker 2014). The "Apply threat and error management principles" competency appears across all of the phases of flight competencies.

The phases of flight competencies use the common ICAO elements and performance criteria structure. For example, the "Perform climb" competency contains the following two elements; the performance criteria for the second element are also shown:

Elements:

4.0. Recognize and manage potential threats and errors.
4.1. Perform standard instrument departure/en-route navigation.
    4.1.1. Complies with departure clearance and procedures.
    4.1.2. Demonstrates terrain awareness.
    4.1.3. Monitors navigation accuracy.
    4.1.4. Adjusts flight to weather and traffic conditions.
    4.1.5. Communicates and coordinates with air traffic control.
    4.1.6. Observes minimum altitudes.
    4.1.7. Selects appropriate level of automation.
    4.1.8. Complies with altimeter setting procedures. (ICAO 2006)

Note how the "Apply threat and error management" competency is integrated within the elements and criteria of the "Perform climb" competency.

In contrast to these entry-level competencies, the EBT framework developed for *recurrent* pilot training in flight simulation and training devices (FSTDs) is more closely aligned to the competence tradition, although the use of behavioral indicators to articulate the competencies reveals an element of the performance tradition. The core competencies for recurrent pilot training are broad statements that mostly focus on the more personal dimensions of competence, such as "workload management" or "leadership and teamwork." The core competencies of recurrent pilot training are:

- application of procedures
- communication
- aircraft flight path management, automation
- aircraft flight path management, manual control
- leadership and teamwork
- problem-solving and decision-making
- situation awareness
- workload management.

As an example of how recurrent pilot training focuses on personal dimensions, Table 7.3 shows the full details of the "Aircraft flight path management, manual control" competency, as they are outlined in the competency text.

Table 7.3   Details of "Aircraft Flight Path Management, Manual Control" Competency

Competency	Competency description	Behavioral indicator
Aircraft flight path management, manual control	Controls the aircraft flight path through manual flight, including appropriate use of flight management system(s) and flight guidance systems	Controls the aircraft manually with accuracy and smoothness as appropriate to the situation Detects deviations from the desired aircraft trajectory and takes appropriate action Contains the aircraft within the normal flight envelope Controls the aircraft safely using only the relationship between aircraft attitude, speed and thrust Manages the flight path to achieve optimum operational performance Maintains the desired flight path during manual flight whilst managing other tasks and distractions Selects appropriate level and mode of flight guidance systems in a timely manner considering phase of flight and workload Effectively monitors flight guidance systems including engagement and automatic mode transitions

Source: International Pilot Training Consortium 2013, Attachment 3.

Air Traffic Control Competencies

The ICAO is not the only regulator to have taken an interest in developing competencies for aviation occupations. While the ICAO is in the midst of developing a competency framework for ATC, expected to come into effect in November 2016, work done in Australia has led to competency frameworks for ATC in that country (see Chapter 3 for a more lengthy discussion of competencies within ATC). In Australia, a competency-based curriculum is mandated for entry-level training. This training forms part of the Australian National Training System (NTS), which uses CBE as its sole model. In 2008, the Organisation for Economic Cooperation and Development estimated that 80% of all occupations in Australia—including some aviation occupations—were covered by the competency-based NTS, making Australia a global leader in CBE. Entry-level training for ATC requires candidates to study for the Diploma of Aviation (Air Traffic Control), a qualification with seven core competencies and one elective. The core competencies are:

1. Apply air traffic control communication procedures and services.
2. Manage human performance and team resources during air traffic control operations.
3. Provide SAR alerting and emergency control.
4. Work professionally in an air traffic control workplace.
5. Operate air traffic control equipment and workstations.
6. Manage traffic flow.
7. Manage situational awareness in the air traffic control environment.
   (Transport & Logistics Industry Skills Council 2013, 6)

Like all competency texts in the NTS, these ATC competencies include components that embody the theory of behavioral objectives as articulated by Mager (1962). That is, there are "elements" (succinct descriptions of observable performance), "performance criteria" (which indicate the level of behavior expected of a competent controller) and a "range statement" (which lists conditions under which the performance should be observed). The Australian NTS competencies also include a cognitive component. This component is broken down into "required knowledge" and "required skills," which are regarded as the foundation of the performances specified in the element and performance criteria statements. The Australian competency format is thus a hybrid model, similar in some ways to the ICAO model used for cabin crew safety training. However, recent changes in the way these important documents are structured in Australia will eliminate much of the cognitive information aspect, leaving the behavioral categories intact.

As an example of the Australian ATC competencies, the "Operate air traffic control equipment and workstations" competency covers the "skills and knowledge required to work as part of a team and to manage and operate air traffic control equipment, systems and workstations associated with providing air traffic services" (from the unit descriptor statement). Seven "elements" are included in this competency:

1. Operate communication equipment.
2. Operate surveillance equipment.
3. Use flight progress strips (FPS) and flight data records (FDR).
4. Operate ancillary equipment.
5. Where provided, use graphic facilities for route.
6. Where provided, use graphic or other facilities for display.
7. Use mandated memory prompts.

Each of the elements has certain "performance criteria." For the first element ("Operate communication equipment"), the performance criteria are:

1.1.   Air–ground communication channels are operated in accordance with standard operating procedure.
1.2.   Selective call and priority channels are operated in accordance with standard operating procedure.
1.3.   Communication equipment is operated in degraded mode.
1.4.   Communication equipment is appropriately configured for tasks.
1.5.   Communication malfunctions are recognized, reported and rectified when able.
1.6.   Communication equipment annunciators are included in scan pattern.
1.7.   Communication equipment is operated within performance limitations.
1.8.   Communication alarms and warning messages are actioned.

The Australian NTS permits "customization" of competencies within certain limits. According to an Australian ATC training specialist interviewed for this book, the ATC occupation provides for customization (ATC4). In the "Operate communication equipment" unit, for instance, training for civil aviation ATCs involves detailed training in the use of the EUROCAT control system, which is used widely in Australia. Although information pertaining to EUROCAT is not specified in the competency, training providers build in training in EUROCAT, and the "Operate communication equipment" competency requires students to achieve competence in the use of this system.

After obtaining their NTS-based Diploma of Aviation (Air Traffic Control), novice air traffic controllers enter a different training regime. Recurrent training for air traffic controllers in Australia is also competency-based, but a more sophisticated, industry-specific model of CBE is used. The competencies under this regime are maintained by the Australian Government Civil Aviation Safety Authority (CASA). This model, which is currently at Version 1.2, also employs elements and performance criteria, but these are situated within an ATC-specific conceptualization of the work based on "fields," which "broadly itemize ATC functional demands" (CASA 2005, 5-1). That is, the analysis is initially directed to functions, and then competencies are differentiated within each field. The ATC fields in this model are:

Field 1: Maintaining situation awareness
Field 2: Executing control actions
Field 3: Communicating
Field 4: Operating facilities (CNS/ATM)
Field 5: Contextual behavior.

The following examples from the CASA manual show a breakdown of the units in one of the fields, one of the units in that field, the element in that unit and the performance criteria in the element:

The field "Maintaining situation awareness" comprises the following units:

• updating traffic picture
• interpreting and evaluating traffic events
• prioritizing, projecting and planning.

The unit "Updating traffic picture" has one element (Scanning), which may be assessed according to performance criteria, which are:

a. never missed critical information
b. may have occasionally suffered from information overload
c. may not always have adjusted rate of scanning to accommodate workload
d. was always able to safely recover from b) and c). (CASA 2005, 5-1)

As mentioned previously, the ICAO is currently in the process of finalizing competencies for ATC, and different regions have also developed ATC competencies. For example, work by Oprins, Burggraaf and van Weerdenburg (2006) produced an ATC occupation-specific model of competency that has a European focus.

Competencies and Instructional Design

The CBE approach has a complex relationship with the discipline of instructional systems design (ISD). According to Smith and Ragan (2005), instructional design is a body of knowledge that covers the areas of:

• learning needs analysis (analysis of problems that may be amenable to training interventions, analysis of potential learners, analysis of contexts of learning)
• learning task or goal analysis
• analysis of evaluation and assessment relevant to any intervention

- design of learning strategies and systems aligned to the results of the prior analyses.

The usual trigger for the application of instructional design processes is the perception that there is a problem in an organization that relates to human capability or performance. The initial needs analysis phase of instructional design comprises frameworks and techniques to investigate the problem, with a view to determining whether a learning intervention is part of a possible solution. This judgment will be informed by an analysis of potential learners and of the context of the problem.

If it turns out that a learning intervention would be likely to help solve the problem, there is a range of theoretical tools available to analyze what needs to be learned. We encountered some of these tools in this chapter when we looked at how data about competent work is analyzed. Smith and Ragan (2005) explicitly conceptualized the goal of a learning intervention in terms of task analysis, and they drew on the conceptual framework of Gagné and colleagues for identifying the kinds of knowledge and processes (e.g., declarative knowledge, procedural knowledge, psycho-motor skills) that underpin the ability to do the task. For Smith and Ragan, the instructional design phases of evaluation/assessment design and learning strategy/systems design must use the results of the learning needs analysis and task analysis to produce a coherent intervention. The alternative framework of Bloom and colleagues was also introduced in this chapter. It offers another way to analyze tasks and a considerable body of research has developed to support it.

In the context of a CBE system, it is clear that there is overlap between the work of creating competency texts and the analytic phases of instructional design. In a sense, the production of competencies begins with the assumption that there is a standing "problem," namely, the initial and continuing training of air cabin crew, air traffic controllers, aircraft maintenance staff and pilots. The performance gap is understood generically as a need for "competence" in different occupational areas, and more-or-less sophisticated task analysis methods are used to represent the learning task in the form of competency texts. Therefore, the implementation of a CBE approach to training means absorbing the initial phases of an instructional design approach into the process of creating competency texts. Some CBE implementations have also produced more-or-less detailed specifications for evaluation and assessment, potentially absorbing this phase of the instructional design process as well. However, CBE implementations tend to have little to say about the strategies and systems of training delivery, preferring to leave this component of instructional design in the hands of training organizations.

In the context of CBE, therefore, one substantial part of the instructional design process must still be undertaken to ensure effective learning: the identification and elaboration of strategies and systems of delivery. In addition, where CBE implementations do not provide detailed evaluation and assessment specifications, a role exists for the development of quality processes and resources to support evaluation and assessment. In Chapters 8 and 9, we will address the facilitation and assessment of CBE. Those chapters will discuss relevant theories and techniques drawn from the instructional design body of knowledge, as well as the more practical questions of how to assess and teach within a CBE system.

However, the process of developing competency statements/texts may leave room for further analysis by instructional designers. In particular, if less sophisticated approaches to competency text design are employed, there may be value in more fine-grained learning needs analysis and/or task analysis to promote effective learning. For example, at the very early phase of design when the potential need for learning is raised, there are questions about the "problem," learners and context that may be overlooked or impossible to address in the design of competencies. It may be appropriate to re-analyze the "problem" that the competencies are expected to solve, in order to determine whether there are learning needs that have not been picked up in the competency design process. The learning task as envisaged by the competency designers may need to be reconceptualized so that a more complete solution can be found. The competency designers' analysis of the learners may not take into account the kinds of learners who might be encountered in particular situations, suggesting another opportunity for further analysis that may improve the targeting and effectiveness of a CBE implementation. Again, the learning context (local resources in addition to organizational, industry and national contexts) may require specific analysis to ensure that a CBE implementation will be effective.

Finally, if the task analysis phase of competency text design has used an unsophisticated framework (such as that associated with Norton's [2004] version of the DACUM technique), there is scope for re-analysis of tasks using more fine-grained schemes such as the one developed by Gagné and colleagues. Competency texts will generally provide a reasonably serviceable description of the learning task or goal of instruction, allowing the instructional designer to determine the types of learning implicit in it. Such an analysis may result in a more rich and useful list of learning outcomes than those included in lists of "knowledge" and "skills" that may be included with statements of elements and performance criteria. A re-analysis of the learning task can generate information that will be helpful for designing learning strategies and systems, as well as for developing evaluation and assessment tools and processes.

For those who are interested in reading more about the theoretical tools of instructional design for the purpose of learning needs (re-)analysis and learning task (re-)analysis in the context of CBE implementations, texts by Smith and Ragan (2005) and Morrison, Ross, Kemp and Kalman (2010) provide comprehensive accounts of these theoretical tools.

The aviation industry has produced a competency framework covering the tasks of instructional design. The *Course Developer* (Develop competency-based training and assessment) framework (ICAO 2006, 5-A-5) includes three competencies:

1. Conduct analysis.
2. Develop training material.
3. Evaluate training material.

These competencies encompass knowledge and skills related to both the design and development of competency texts and the design of training strategies, assessment and evaluation. It is a framework that reflects the traditional expertise of instructional design. It should be noted that although this *Course Developer* framework includes a competency for task analysis, course development in a CBE context means that developers must work with competency texts that already contain a task analysis.

Chapter 8
Training

Questions Addressed in this Chapter:

- How are competencies used in training?
- What are some key issues in learning design for CBE?
- What are some approaches to creating effective programs and sessions for CBE?
- How should e-learning be used in competency-based education?
- How can the quality of e-learning be assessed?

Competency-Based Education

Competency-based education (CBE) is ultimately about teaching people to be competent in their work roles. This chapter focuses on the "training" component of CBE. Although a CBE approach does not dictate particular training methods, it can encourage certain practices. For instance, CBE incorporates "application" as part of the training process (rather than expecting graduates to learn how to apply knowledge, skills and attitudes independently on the job), and therefore training methods that integrate realistic opportunities to practice new learning are appropriate. CBE also encourages methods that allow students who learn faster to graduate faster. By the same token, if students are having difficulty learning through the techniques favored by their trainer, the flexibility inherent in the concept of CBE gives them extra opportunities to succeed through alternative training materials. Despite these possible advantages, organizations and instructors may struggle to create curriculums that benefit from the realistic focus and flexibility of CBE. For example, if efforts to put innovative strategies in place are undermined by a lack of understanding of competency-based teaching methods or assessment, or if there are resource restraints, traditional methods may prevail. This may result in learning that takes place in settings that are not representative of the real-world job, or training may be implemented in a "lock-step" fashion, with all students expected to learn at the same pace.

How can training managers and instructors take advantage of the opportunities offered by competency-based methods? This chapter describes the more innovative training potential of CBE, starting with a consideration of the principles of training strategy design derived from the instructional design literature. The chapter will move to some practical guidelines for conducting training, before exploring e-learning as an increasingly important approach to training that is widely used in

CBE. But first, the use of competency statements/texts will be considered, since these texts are the central reference point of the CBE approach.

The Role of Competency Statements/Texts in Training

The previous chapter described ways in which competency statements/texts, or "competencies," are designed. It was suggested that this process mirrors the early steps in the instructional design process, beginning with a training needs analysis, specification of the tasks that will need to be learned and moving on to some form of analysis of the task to highlight the types of learning implicit in the task. The result of the competency design process is a concise description of the task that needs to be learned and any components (such as particular knowledge and skills) of the task that may directly inform training strategy design. The previous chapter also suggested that the process of writing "competencies" may not produce a sufficiently complete, contextualized or detailed specification, and that additional instructional design work may be needed to create an appropriate starting point for training. Overall, however, the fact remains that in CBE, **competencies must be the key reference point for the design of training**, even if additional information may be gathered to complement the information in the competency text. Three basic questions must be addressed in relation to the use of competencies for training: questions about how to use, combine and understand competencies and competency texts. These three questions are presented and discussed below.

1. How Are Competencies Supposed to be Used within Training?

Competency texts describe learning outcomes. As explained already in this book, the structure of competencies may reflect the components of the classic "behavioral objective": a description of the desired performance, an indication of the level of performance expected and some statement of the conditions under which the performance must take place. In the aviation industry, competencies are often formatted so that the performance description component is captured in statements of "elements" and the level of performance is described in "performance criteria," while conditions may or may not be specified separately. Competencies with this kind of design present a basic description of the learning task across its elements, while the performance criteria help to further specify the learning outcome (see Figure 8.1).

Most training design approaches make extensive use of statements of learning outcomes. For instance, content from competencies may be used to inform learners of what is required, learning activities can be devised to achieve the task described in the competency and assessment (as we will see in the next chapter) can be designed to measure the extent to which learners can perform the task described in the competency. In this case, the competency texts underpin a coherent approach to training and assessment that potentially keeps learners fully informed about learning expectations from the start.

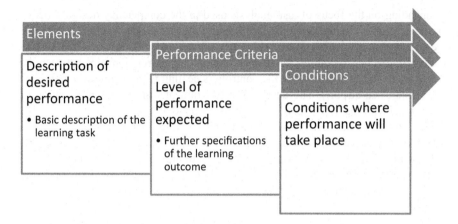

Figure 8.1   Relationship between Behavioral Objective and Competency Texts

2. How Can Competencies be Combined to Form a Curriculum?

Competencies are often conceptualized in sets, with multiple complementary competencies designed to address a more complex task or work role. Rarely are single competencies designed to function as stand-alone bases for training. In a CBE system, the training designer, trainer and assessor are usually required to work with multiple related competencies. This aspect of work in the CBE system can be quite challenging, particularly if funding, credentialing and auditing regimes are calibrated to the individual competency level. In addition, there is often little explicit guidance for combining competencies—that is, which particular competencies to combine and how to combine them. However, sets of competencies are often designed to articulate whole complex tasks or roles, and it may be clear how particular competencies are supposed to relate to each other. Individual industries may offer directions for combining competencies, and sometimes competencies are presented in groups, as they are within an ICAO "competency framework." Industries may designate certain competencies, or sets of competencies, as prerequisites for training and assessment in other competencies. So, how competencies are combined to form a training curriculum can be determined either by the training designer (based on their expertise) in conjunction with their organization (such as the chief pilot or head of training and standards) or by the designers and administrators of the competencies. Either way, there must be careful consideration of which competencies relate to one another before training design begins.

A problem for training designers who are required to address multiple competencies is whether to approach the content of the texts sequentially or to integrate the content in a holistic design (or some combination of these approaches).

Depending on the focus of task analysis used in the competency design process, sequences of competencies may constitute a faithful reflection of an overall role (e.g., emergency opening up of an aircraft door during a passenger evacuation). In other approaches to competency design, more complex relationships between the texts are envisaged (e.g., air traffic controllers dealing with an aircraft emergency at a congested airport). The design of a program of instruction and creation of individual learning activities therefore must respond appropriately to the nature of the *relationships* between related competencies. For instance, where competencies are related in a sequential fashion—as in door operations—it may be appropriate to develop a series of workshops, each devoted to individual competencies. When learners have completed the series of workshops, they will have developed the sequence of competencies that reflects an actual role or complex task used in the industry.

On the other hand, where competencies are related in a more complex way, a larger holistic learning task or project may be appropriate, which encourages learners to integrate performances described in individual competencies. For example, after a group of air traffic control students have been given an opportunity to deal with an aircraft emergency on departure, ample time would be given to a group debriefing, where all those involved would participate. Here the controllers would collectively outline what had happened, things that went well and not so well, and strategies to improve future practice. The senior trainer's role would be that of a facilitator. The senior instructor would give direct instruction only if the group was unable to come up with appropriate solutions or if the solution was incorrect. On completion of the debrief, the senior instructor summarizes the session, key learning points and, more specifically, areas that each individual within the team needs to work on. One controller may require additional study focusing on technical knowledge of separation. Another may need to focus more on how to work more efficiently with other team members during the emergency itself. In summary, the holistic training session focuses not only on the sequential use of single competencies, but also on the complex relationship between competencies.

3. How Are Competency Texts Understood by Trainers?

Successful implementation of CBE assumes that designers, trainers and assessors (and perhaps learners too) can readily decode the information contained in competency texts. However, no matter how much attention is given to the clear expression of competency texts, individual stakeholders (training designers, trainers, etc.) may not interpret the competency in the way intended by the competency writers. Research by Hodge (2014) indicated that trainers in the Australian competency-based vocational education sector did not always understand the terse style of writing used in the texts. They also did not always understand the nature of different categories of information (elements, performance criteria, etc.). The trainers interviewed for this book also tended to adopt a limited strategy for interpreting competencies, which involved focusing on one or more components of the texts. The use of holistic strategies—that is, seeking to understand all the components of the texts and how they interrelate—was a relatively rare approach. In addition, our interviewees generally did not attempt to have their interpretations validated by other trainers or industry experts. The authors of this book found a similar phenomenon among aviation professionals: rather than trying to have their interpretations of competency texts validated, there was a tendency to promote their individual understanding to others.

Hodge (2014) argued that work within a CBE system entailed a "hermeneutic" form of expertise (meaning expertise in the consistent and rich interpretation of competency texts) in addition to the requisite occupational expertise. The term *hermeneutics* was originally used to describe the systematic interpretation of scriptures. According to Hodge, CBE trainers need hermeneutic skills because the interpretation process is so important in a text-based system. Hodge recommended that training in this skill be considered a priority in implementations of CBE and that trainers should be encouraged to share and validate their interpretations. When the hermeneutic dimension of CBE implementation is not taken into account, a vulnerable link in the implementation—the use of competencies by people other than the competency writers—becomes fraught with risk, as differences in individual interpretations may vary widely.

Learning Design

Instructional design research and theory have produced a wealth of guidance for training designers, trainers and assessors. In this section we briefly consider the alignment of training strategies for competency-based learning, the distinction between supplantive and generative approaches to creating training, and some templates for training design. These templates can be used for individual training sessions or as building blocks for whole programs.

Alignment of Strategies and Types of Learning

Instructional design research and theory make suggestions about what training strategies should be used to accomplish particular kinds of learning outcomes. Different kinds of activities promote different kinds of learning. Tapping into Smith and Ragan's (2005) comprehensive overview, Table 8.1 briefly describes strategies appropriate to different kinds of learning outcomes (based on Gagné and colleagues' analysis of types of learning). The labels for different kinds of learning—declarative, discriminating, conceptual, etc.—in the first column are accompanied by a translation into more basic designations (in parentheses). The second column provides a very brief description of a strategy appropriate to developing each kind of learning.

Table 8.1   Example Strategies for Gagné et al.'s Learning Outcomes

Learning outcomes	Example strategies
Declarative (Knowledge)	• Associational techniques: use of mnemonics, images and analogies • Organizational techniques: clustering and chunking by categories, using graphic organizers, generating expository and narrative structures, and using advance organizers • Elaborative techniques: making elaborations on the material being learned, including elaboration into sentences and devising rules
Discriminations (Knowledge)	• Ask learners to decide between options that represent the differences being learned, with feedback provided on decisions
Concepts (Knowledge)	• Inquiry strategy: present examples and non-examples of the concept and prompt learner to induce or "discover" the concept underlying the instances
Principles (Knowledge)	• Present learners with a puzzling situation that shows the relationship between the variables: learners question the trainer about the situation, but can only be answered in yes/no terms until they have discovered the rule/principle that connects the variables
Procedures (Skills)	• Presentation of the procedure, with demonstrations of the application of the procedure
Problem-solving (Skills)	• Prerequisites of problem-solving are relevant principles, associated concepts, declarative knowledge and cognitive strategies. Case studies allow the development of problem solutions, assuming that learners have the prerequisite knowledge
Cognitive strategies (Skills)	• Invite learners to analyze the requirements of the learning task, analyze their ability to complete the task, select an appropriate strategy, apply the selected strategy, evaluate the effectiveness of the strategy used and revise as required
Attitudes (Attitudes)	• Role-modeling is regarded as an essential component in any attitude learning, while role-playing allows learners to assume and practice desired attitudes
Psychomotor skills (Skills)	• Demonstration and practice of muscular movements, combined with learning procedural rules relevant to the skill

Source: Smith and Ragan 2005.

In the last chapter Bloom et al.'s taxonomies were described. They offer an alternative to Gagné et al.'s approach to breaking down learning types. Researchers have identified a range of strategies aligned to Bloom's taxonomy which can assist in the development of effective training.

Generative versus Supplantive Types of Training

Training must be well calibrated to learners' familiarity with the content. If learners know relatively little about the content, training needs to supply new material and rely less on prior learning as a platform. If learners have some relevant expertise or experience in the content, active approaches that challenge learners to connect existing knowledge with new material can be used. "Generative" strategies may be used for learners with prior knowledge in the area, while "supplantive" strategies are more appropriate for novices in the area (Jonassen 1985; Smith and Ragan 2005).

Supplantive strategies are appropriate when:

- learners have little prior knowledge of the content
- content is critically important
- content is likely to place a high cognitive load on learners
- less time is available for training (Smith and Ragan 2005).

In contrast, generative approaches are appropriate when:

- learners have some prior knowledge of the content
- content is lower risk
- learners can cope with high cognitive load
- more time is available for learning (Smith and Ragan 2005).

Supplantive Approach	Example: 1-hour session in a classroom with an instructor using a PowerPoint presentation

- Interaction: Question-and-answers with learners
- Goal: learners emerge with new knowledge that can be applied
- Note: Learners without background knowledge may feel overwhelmed by the unstructured approach

Generative Approach	Example: 3-hour scenario-based exercise led by the instructor

- Interaction: Trainer provides less direct guidance, and learners interact and problem-solve
- Goal: learners emerge with a good sense of how new material relates to their background knowledge and experience

Figure 8.2   Examples of Supplantive and Generative Approaches

The concept of "cognitive load" is important here. Generative training methods that encourage exploration, problem-solving and discovery of solutions may be ineffective if cognitive load on learners is too high. The decision to choose a generative versus a supplantive strategy is a matter of finding a balance (Smith and Ragan 2005). In more extended training programs, it may be appropriate to use supplantive strategies early in the program and move toward more generative strategies as the program unfolds.

Templates for Training Design

Events of instruction model. Researchers have proposed a range of templates for promoting learning, whether they be of a supplantive, generative or combined type. Gagné's (1970) "events of instruction" model is a template derived from an analysis of human cognition, based on information processing theory. The events of instruction model can be used as a guide to suggest how certain learning activities in a particular sequence are likely to optimize learning. Gagné's events of instruction model contains the following nine events:

1. *Gaining and controlling attention.* The trainer gets the learners' attention.
2. *Informing the learner of the expected outcomes.* The trainer informs learners of the expected outcomes of the training.
3. *Stimulating recall of relevant prerequisite capabilities.* The trainer reminds learners of relevant knowledge and skills they have already learned.
4. *Presenting stimuli inherent to the learning task.* The trainer explains and demonstrates what the learners will need to do to demonstrate achievement.
5. *Offering guidance for learning.* The trainer gives prompts and hints to guide learners to achievement.
6. *Providing feedback.* The trainer informs learners of how correctly they are performing the task.
7. *Appraising performance.* Learners are given opportunities to practice and refine their performance.
8. *Making provisions for transferability.* Learners are given extra examples of application of their learning.
9. *Insuring retention.* Learners engage in extra practice to enhance retention (adapted from Gagné 1970, 304).

Smith and Ragan (2005) built upon Gagné's events of instruction model by addressing the two main forms of learning design (supplantive and generative). Their 15-step version, called the expanded events of instruction model, is represented in Table 8.2.

Table 8.2   Expanded Events of Instruction Model

Expanded events of instruction	
Generative ... student generates	*Supplantive ... instructor supplies*
Introduction	
Activate attention to activity	Gain attention to the learning activity
Establish purpose	Inform learner of purpose
Arouse interest and motivation	Stimulate learner's attention/motivation
Preview learning activity	Provide overview
Body	
Recall relevant prior knowledge	Stimulate recall of prior knowledge
Process information and examples	Present information and examples
Focus attention	Gain and direct attention
Employ learning strategies	Guide or prompt use of learning strategies
Practice	Provide for and guide practice
Evaluate feedback	Provide feedback
Conclusion	
Summarize and review	Provide summary and review
Transfer learning	Enhance transfer
Remotivate and cease	Provide remotivation and closure
Assessment	
Assess learning	Conduct assessment
Evaluate feedback	Provide feedback and remediation

Source: Smith and Ragan 2005, 130.

Smith and Ragan's (2005) template is designed to allow generative and supplantive versions of the events of instruction approach pioneered by Gagné (1970). Within the generative approach, the trainer has more of a facilitative role, while the learner takes the lead. In the supplantive approach, the trainer and training organization directly resource and guide the events, and learners receive instruction led by the trainer.

Whole-task training. In terms of the principles and potential of CBE, either supplantive or generative approaches can be used or combined, depending on factors mentioned above (e.g., level of prior knowledge possessed by learners, time available for training). However, a focus on practice is central to CBE, given that *application* of newly trained knowledge and skills is of key importance. A learning design that is highly appropriate to CBE is the "whole task" method. This approach would sidestep the disadvantages of CBE associated with the fragmentation of slicing a complex real-world workplace into artificially simplistic competencies that ignore connections between tasks and their underlying meaning. Oversimplification can result in CBE that produces learners who demonstrate a "checklist" of knowledge, skills and attitudes without ever truly excelling or critically thinking about real-world scenarios (Leung 2002).

However, whole-task instruction is also far from foolproof. A limitation of whole-task instruction is that novices may become overwhelmed by information (i.e., not cope with high cognitive load). This can be moderated by designing learning activities that begin simple (i.e., in a supplantive mode) and then become increasingly complex while providing just-in-time information to support learners (including information required to support practice, as well as step-by-step explanations of procedures; van Merriënboer, Kirschner and Kester 2003).

The 4C/ID model (van Merriënboer 1997) supports a CBE approach by designing learning environments that allow learners to develop complex knowledge, skill, attitude and application competencies. Training is organized into a preliminary unit of supportive information, explaining foundational information about the topic, followed by several units of whole-task experiences that range from artificially simple to complex. Just-in-time information (such as books, checklists and computer-based tools) and, when necessary, repetitive part-task practice are integrated (van Merriënboer, Clark and de Croock 2002).

In whole-task training a learner is required to draw information from a pool of knowledge, skills and attitudes to apply to a real-world scenario. The formal distinctions between course topics are not relevant. The philosophy is that the more separate and isolated knowledge, skills and attitude are during training, the longer it takes to teach the time-sharing/management skills to link these attributes back together for real-world performance.

Whole-task experiences may take a range of forms, including case method analysis, e-learning, high- or low-fidelity simulation, apprenticeship, job-shadowing and analysis, classroom tutorials or even strategic games. The important considerations are that the instruction is based on an appropriate design strategy (see Table 8.3), the learner is actively engaged in the process and the task targets an authentic, realistic, whole-task scenario.

Table 8.3   Whole-Task Design Strategies (adapted from Paas and van Merriënboer 1994)

Whole-task design strategy	Description	Example	Results
Hierarchical	The knowledge required for a "whole task" is broken down into a hierarchy (primary objectives, optimal strategies for each, identification of subskills). Learners study skill and knowledge components separately and apply them to the whole task at the end of training (Frederiksen and White 1989). The downside is fractionation—when two or more tasks are performed simultaneously in the "whole task," students can miss the chance to practice important time-sharing skills (like scanning, task switching, etc.; Wickens, Hutchins, Carolan and Cumming 2013).	Managing an emergency in the flight deck requires a high level of awareness and synchronization of captain and first officer duties. During early training, pilots not provided with templates to follow find these tasks difficult. To assist, pilots are taught to break the process of managing an emergency down into (1) flying the airplane, (2) identification of malfunction, (3) securing the malfunction (e.g., through use of a checklist), (4) identifying implications of malfunction to flight and (5) possible contingency plans. These skills are taught separately and then later combined.	Learners show better transfer compared with those who had to perform the whole task from the beginning of training (Fabiani et al. 1989). Fractionation must be avoided because it results in increased training times, as learners need time to develop time-sharing skills that were missed (Wickens et al. 2013).
Variable priority/ emphasis manipulation	Learners complete the "whole task" intact; however, the instructor varies the difficulty/complexity of subcomponents (Gopher, Weil and Siegel 1989; Wickens et al. 2013). Allows an instructor to lower the level of difficulty without sacrificing the integrity of the "whole task" training experience. As learners progress in training, difficulty is increased until it reaches the operational level.	During early training in air traffic control, students will be required to control only a small number of aircraft. As student abilities improve, aircraft numbers will increase, as well as type, such as faster twin-engine airplane, larger commercial regular public transport aircraft and aircraft operating under visual or flight instrument flight rules.	Allows lower-ability students to achieve the same results as high-ability students, as they can pay attention to smaller subtasks and avoid cognitive overload. May allow learner to develop a broader perspective of task and result in more flexible decision-making.

Goal-free or nonspecific goal	Instead of giving learners a clear problem with a "correct" answer, they are challenged to think creatively to solve a problem that is not clearly defined. For example, giving learners a scenario and asking them to think of several actions and the resulting outcomes of each. This strategy shifts a learner's focus away from identifying a quick solution toward a strategy where they must work forward from the given information.	During cabin crew training, possible implications of disruptive, disorderly and violent passengers are discussed. Here the instructor will ask students to come up with possible ways to differentiate the three levels, possible behaviors passengers may display and different ways that they may deal with each of these possible scenarios.	Enhances learning more than practice with traditional problems, reduces cognitive load, results in higher transfer of skills (Owen and Sweller 1985).
Worked-out problems/worked examples	Learners are given a written problem with an expert's well-structured solution (providing the best example). The goal is to make the "invisible" expert thinking and problem-solving process visible to learners so that they may extract problem-solving principles. It is very important that all problems be relevant to the real-world environment. As skill increases, learners will be given only part of the solution and are tasked with completing the solution (Van Gog, Paas and van Merriënboer 2004).	Some examples of these have included major aircraft accidents where the outcome—even given the extreme circumstances—is very good. Examples include US Airways Flight 1549, where Captain Sullenberger and First Officer Skiles successfully landed a stricken A320 into the Hudson River. Post-interviews of both flight crew provided descriptions of their thinking prior to and during the event.	Can optimize transfer of training, can improve problem-solving skills more than actually solving a problem (Sweller and Cooper 1985). Avoids cognitive overload by allowing learners to take the time needed to review the problem and model the expert solution (Sweller 1988).

E-Learning and CBE

Another consideration in the creation of CBE is the method of delivery. Although classroom, simulator (further discussed in Chapter 10) and on-the-job training are still used, e-learning is an increasingly popular instructional medium. E-learning can be a particularly attractive option within competency-based instruction, as it can support a customized and flexible approach to training, which is well aligned with CBE. Although CBE standardizes the goal of instruction among learners in a professional group to a predetermined competency text, the training material given to each learner can be customized to individual learning needs and background experience, and to the training requirements of the unique job task. E-learning can support this process to avoid a cookie-cutter approach to training that may not represent real-world challenges or what the learner needs to know.

What is E-Learning?

E-learning is a shortened version of the term *electronic learning* and refers to educational materials, computer-mediated communication and the delivery of instructional content through technology (Eddy and Tannenbaum 2003). The focus of this approach should remain on the learning rather than on the technology used to deliver it. E-learning began with military training videos in World War II and evolved with the invention of the personal computer in the late 1980s (Rosenberg 2001). This early e-learning consisted of CD-ROMs that would install training programs on the learner's computer. The aviation industry was early in recognizing the advantages of e-learning, and aircraft manufacturers began to develop proprietary learning programs (Kearns 2010). Early computer-based training was plagued with difficulties, as courseware would function only on certain computer platforms, limitations of software made training boring (mostly long readings followed by multiple choice questions), development costs were very high and a lack of understanding of instructional design resulted in the creation of many ineffective courses (Rosenberg 2001).

The Internet changed e-learning by allowing for web-distributed content that can be accessed on demand. The Internet has changed how an entire generation of people access information. The next generations of aviation professionals are likely to be more comfortable referencing electronic information rather than paper-based sources (Kearns 2010). Modern e-learning is widely utilized and evidence suggests that aviation professionals are capable of learning through this method and are comfortable with new technologies (Raisinghani et al. 2005).

E-learning can be organized into three broad categories related to the role of technology in connecting instructors and learners: synchronous, asynchronous and blended learning (Kearns 2010).

- **Synchronous.** Learners and instructors log in to an online classroom simultaneously at a specified time from their respective locations. The pace of the curriculum is controlled by the instructor. This approach typically involves computer-mediated communication, such as instant messaging software or a webcast using a webcam and headset.
- **Asynchronous.** Learners complete training independently, without a simultaneous connection with an instructor. Some asynchronous courses are instructor-led, while others are self-instructional, meaning that students work through the material independent from a live human instructor.

If instructor-led, communication between learners and the instructor is time-delayed and typically takes the form of a message board post that others can read and respond to at their convenience. In a self-instructional course, learners progress through courseware at their own pace. Self-instructional asynchronous courses are regarded as particularly advantageous within aviation training institutions because the courseware is *scalable*—meaning it can be delivered to one learner or to 1 million learners with minimal additional costs (Kearns 2010).

- **Blended learning.** This is the combination of two types of learning environments to achieve an outcome. The foundation of aviation training, combining both practical instruction in a real-world context with classroom theoretical training, is an example of blended learning. A combination of electronic courseware (synchronous or asynchronous) with instruction delivered in a traditional classroom—referred to as a "hybrid" course—also represents a blended learning approach (Kearns 2010). Within a blended course, electronic courseware can be used to support, rather than replace, classroom instruction. Most colleges and universities use this model, in which web-based learning management systems (LMSs) support classroom teaching through a portal that learners and instructors use to distribute messages, media, grades and other classroom materials.

At a time when our industry is increasingly replacing classroom instruction with more economical e-learning, it is important to consider a word of caution. There are many software packages on the market that allow organizations to quickly convert classroom courses into e-learning courseware. These programs, which are considered "rapid repurposing" tools, typically cut-and-paste classroom content into an online platform, and the end result is an asynchronous web-based slideshow that sometimes includes an instructor voiceover. Although this accomplishes the goal of creating online courseware, no consideration is given to the quality of the content or of delivery. This kind of online course is unlikely to have the instructional design, interactive exercises or assessments that are required to be effective. This approach eliminates all of the interactions, discussions and activities that would be facilitated by a skilled instructor if the material had been delivered in a classroom (Kearns 2010). Designing effective e-learning requires expertise, imagination and time—all in large amounts (Piskurich 2006). This concept will be explored in additional detail throughout the rest of this chapter.

Advantages and Disadvantages of E-Learning

E-learning offers several advantages, including:

- cost efficiency
- the ability to deliver training to an unlimited number of learners

- flexibility of time and location of training
- standardized content among instructors (to avoid the variable success of a course resulting from "good" and "poor" instructors within the same organization)
- interactive activities
- immediate feedback
- performance tracking (Kearns 2010).

However, there are also important disadvantages:

- the creation and design of courses may cost more than projected savings
- learners must be more self-directed
- direct face-to-face contact with instructors is lost
- the entire structure of training must be rethought (it cannot be built directly from classroom slides but must be created with electronic delivery in mind) (Kearns 2010).

If an e-learning course is developed from the exact same materials as a classroom-based course, at best it will result in similar learning outcomes to the original face-to-face course (Kearns 2010). However, the true effectiveness of e-learning compared with classroom instruction is polar: some e-learning courses perform much better, and others much worse, than classroom courses. To ensure that an investment made in e-learning is not wasted, it is crucial to create e-learning based on sound instructional principles. A significant body of work indicates that e-learning can result in better learning outcomes because it has characteristics such as the following:

- Courses are built on instructional design practices that capitalize on the capabilities of learning (such as relevant interactivity and immediate feedback; Bernard et al. 2004).
- Blended learning, which incorporates both e-learning and classroom instruction, results in better learning outcomes than either method alone (Cavanaugh 2001).
- Courses that allow for learner contact with live human instructors (through synchronous courseware or face-to-face instruction) are more effective than those where students interact solely with technology (Zhao et al. 2005).
- E-learning is effective at developing declarative knowledge (such as foundational knowledge "of" a topic; Norman 2002; Sitzmann et al. 2006).

However, some characteristics result in less effective e-learning courses:

- Rapidly developed e-learning courses with content that has been quickly adapted from the classroom tend to significantly underperform their classroom equivalents (Cavanaugh 2001).

- E-learning courseware that targets procedural knowledge, such as practical real-world knowledge associated with how to "do" something, is less effective than courseware that targets declarative knowledge (knowledge "of" something; Sitzmann et al. 2006).

Beyond these broad methodological characteristics, research has been conducted comparing courses with similar content but different multimedia elements (pictures, text, videos, etc.). Clark and Mayer (2011) organized the findings of this body of work into eight key media principles that have been found to improve learning outcomes. These strategies can be summarized as follows (adapted from Clark and Mayer 2011):

1. Use both pictures and words.
2. Keep pictures and words that are related close together.
3. Use audio instead of written text whenever possible.
4. Explain pictures with an auditory voiceover or onscreen text, not both.
   - Never have the audio voiceover read the exact text that is written on screen. The voiceover should mimic the role of an instructor, guiding the learner through the content and offering insight and explanations, rather than dictating onscreen text.
5. Eliminate material that does not relate to training objectives.
6. Relate to trainees by using a friendly conversational tone rather than the kind of technical language that would be used in a textbook.
7. Incorporate friendly digital onscreen coach characters.
8. Break training down into manageable chunks.

How to "Blend"—Better than Classroom or E-Learning Alone

As mentioned already, learning outcomes are better in courses that include both classroom and e-learning components. This may lead to questions around what training material is better in a classroom and what should be delivered electronically. The decision strategy regarding how to incorporate e-learning within a training curriculum should be based on what material is enhanced when delivered in a face-to-face modality. Face-to-face may be preferred when the content is best understood through an experience or interaction, when special equipment such as simulators are required or when back-and-forth discussion with an instructor is needed for complex material.

Once this is determined, other training material can be segregated and considered for an e-learning platform. Suitable content areas may include background knowledge completed before a learner arrives for training or homework completed in the evenings after face-to-face sessions. E-learning is also a useful tool to extend the training after the learner has left the training center.

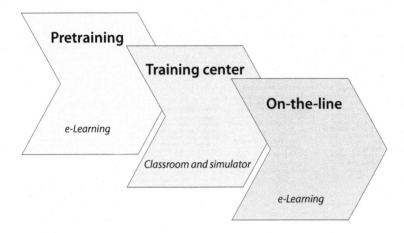

Figure 8.3   A Blended Learning Strategy for Aviation Training
Source: Kearns 2010.

This approach can be visualized through the sandwich model presented in Figure 8.3, which shows how face-to-face instruction can be sandwiched between e-learning.

When competency-based education is the foundation for training conducted at the training center, there are additional applications for the use of e-learning. E-learning can be developed as a pool of electronic resources that are available to learners as they work through problem-based exercises. The important consideration is to ensure that the instructional objective is aligned with a competency and then create training materials to accomplish the objective, incorporating opportunities for learners to practice applying new material to job-related tasks (such as through whole-task training, discussed previously in this chapter). E-learning can be a practical tool to support this approach by providing foundational information and just-in-time reference materials that learners can reference when necessary.

Short units of training, such as "snap-courses" of five to seven minutes each, are one e-learning instructional strategy that can be used (Kearns 2013b). Snap-courses are designed to be components of a larger training curriculum. They incorporate the following characteristics:

- five to seven minutes in length
- interactive
- designed for personalization, rewards and choice to improve learner motivation
- integrated quizzes.

Snap-courses support competency-based education approaches and can be regarded as pieces of a jigsaw puzzle. Snap-course pieces can be combined to create a curriculum that corresponds to the unique needs of each individual.

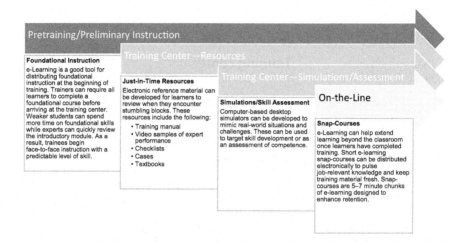

Pretraining/Preliminary Instruction

Foundational Instruction
e-Learning is a good tool for distributing foundational instruction at the beginning of training. Trainers can require all learners to complete a foundational course before arriving at the training center. Weaker students can spend more time on foundational skills while experts can quickly review the introductory module. As a result, trainees begin face-to-face instruction with a predictable level of skill.

Training Center—Resources

Just-In-Time Resources
Electronic reference material can be developed for learners to review when they encounter stumbling blocks. These resources include the following:

• Training manual
• Video samples of expert performance
• Checklists
• Cases
• Textbooks

Training Center—Simulations/Assessment

Simulations/Skill Assessment
Computer-based desktop simulators can be developed to mimic real-world situations and challenges. These can be used to target skill development or as an assessment of competence.

On-the-Line

Snap-Courses
e-Learning can help extend learning beyond the classroom once learners have completed training. Short e-learning snap-courses can be distributed electronically to pulse job-relevant knowledge and keep training material fresh. Snap-courses are 5–7 minute chunks of e-learning designed to enhance retention.

Figure 8.4   Integration of E-Learning within Various Phases of CBE

This is a departure from traditional e-learning courseware that is typically two to three hours in length and requires learners to view the entire course regardless of their pre-existing knowledge or training requirements. With a snap-course approach, e-learning can support learners by providing a rich and in-depth resource for learners who are struggling, while allowing advanced learners to demonstrate competence with a very short time investment.

Levels of E-Learning

Unfortunately, most industry decisions regarding the integration of e-learning are based on the projected cost savings of using e-learning to replace some of the classroom teaching. As a result, many organizations request proposals from e-learning developers and typically choose the least expensive proposal. A problem with this approach is that there is a wide variability in the quality and effectiveness of e-learning courses. Digital instruction ranges from simple page-turning courses that are little more than digital textbooks, to advanced scenario-based simulators that realistically replicate real-world challenges. To provide some insight to decision-makers who are comparing different e-learning offerings, it is helpful to consider different "levels" of e-learning (Chapman 2010). The four levels of e-learning are associated with increasing amounts of media usage, interactivity and sophistication, and their development costs vary significantly as well. Table 8.4 presents some industry averages regarding the cost and development hours for various levels of e-learning, ranging from simplistic to highly complex.

Table 8.4   Four Levels of E-Learning

Level of e-learning	How is it made?	Average development time (for 1 hr of training; Chapman 2010)	Average development cost (for 1 hr of training)	What does it look like?
Classroom	Instructor-led and developed	40 hours	US$6,000	**Watching.** Learners typically sit and observe the instructor's lesson.
1st-level elearning	Instructional content is repurposed from classroombased training	80 hours	US$10,000	**Watching.** Typically PowerPoint slides or a videotaped classroom session, learners occasionally click the "Next" button. Training followed by an assessment
2nd-level elearning	Instructional content is repurposed from classroom-based training with added interactivity	200 hours	US$18,500	**Watching & clicking.** Resembles a PowerPoint presentation, but includes: • media (videos, graphics, animations) • interactive exercises (drag and drop, quizzes). Training followed by an assessment.
3rd-level elearning	Instructional content is custom designed for a digital environment	500 hours	US$50,000	**Immersive & interactive.** User experience in each course varies (may include advanced learning simulations and games): • Interactions are customized and require more complex thought from learners • Presented by avatars (cartoon characters) using a conversational tone. Training is delivered in short chunks (no more than 15 minutes at a time). Training followed by an assessment.
4th-level elearning	Custom instructional content is embedded within an intelligent computerbased platform that tracks learner strengths and weaknesses, and pieces together a unique learning path	600–1,800 hours	US$100,000+	**Immersive, interactive and individual.** User experience in each course is customized. Courses are not completed linearly (1, 2, 3, etc.). Each learner's training path is customized based on his or her demonstrated strengths and weaknesses: • Expert learners progress quickly, average performers progress toward mastery, and novices receive additional support as required. Assessments are integrated within training and completed continuously to inform the adaptive computer platform.

Sources: Chapman 2010; Kearns 2013a, 2014.

As is evident in Table 8.4, initial development costs for e-learning are generally higher than those for equivalent classroom courses. In addition, costs range dramatically, according to the level of e-learning that is developed. It is helpful to note that Level 4 e-learning is very rare but is expected to become increasingly popular as technology evolves. Level 4 e-learning is discussed in more detail in Chapter 10.

Initial development costs for e-learning courseware can be an expensive undertaking. Yet many organizations fail to realize that initial costs do not account for maintenance/update costs, which can be as much as 30% of the initial development cost each year (depending upon the amount and frequency of content updates). Despite high development costs, financial savings often do materialize, as asynchronous e-learning courseware can be scaled for delivery to an almost unlimited number of learners through online delivery.

If online courses are strategically chosen to be based on content areas with a massive number of students, and on topic areas that do not change much over time, this approach can save organizations a great deal of money without sacrificing the quality of training. For example, TransAsia Airways published their company's costs associated with e-learning development. They invested US$400,000 in the development of e-learning. Afterwards, on an annual basis, they saved US$340,000 on flight crew salaries, US$20,000 on instructor salaries, and US$840,000 in operational savings resulting from the ability to schedule additional flights. This represented a 200% return on investment for the first year—meaning that the airline saved $2 for every $1 spent on e-learning—with similar operational savings recurring every year thereafter (Chuang et al. 2008).

The TransAsia example is a best-case scenario, as it does not account for the effectiveness of training. Clearly, these financial savings offer a convincing argument for the replacement of classroom training with e-learning. However, if the e-learning courses are ineffective, there may be significant losses associated with retraining in the classroom.

ICAO Trainer Competencies

Considering the training design and delivery strategies discussed thus far, it is interesting to consider that a competency framework for aviation trainers has been developed. These competencies align with key concepts discussed in this chapter and highlight other areas we have not considered. The four instructor/deliverer competency-based education competencies are:

1. Prepare for delivery of training.
2. Conduct competency-based training module.
3. Evaluate trainee's performance.
4. Prepare course delivery report (ICAO 2006, 5-A-1).

The first of these competencies (Prepare for delivery of training) has an administrative focus. Some of its elements cover diagnostic assessment (covered in the next chapter). This competency emphasizes the importance of systematic preparation to effective training practice. The second competency (Conduct competency-based training module) includes 22 elements that suggest an "events of instruction" model of training, presented in group and individual instruction versions. Training strategies are included, such as written exercises, group discussion and role play, and e-learning is also identified as an option. This competency appears to authorize a limited range of learning design and training strategy options.

The third competency (Evaluate trainee's performance) makes reference to formative and summative assessment processes that are examined in Chapter 9 of this book. The final competency (Prepare course delivery report) also has an administrative focus. Its elements cover summarization of summative assessment results, provision of feedback to course designers and a summary of trainees' feedback. The administrative focus of this and the first competency underline a feature of CBE highlighted by trainers in other industries—CBE tends to increase the amount of "paperwork" involved in training.

Conclusion

Previous chapters in this book have explored what "competencies" are and how they are created. Yet a crucially important consideration is how competencies can be used to inform competency-based education (CBE). If competencies are used poorly, trainers could form a checklist from them to prove that instruction was delivered on each of the required competencies (a new version of the traditional "check-the-box" approach). However, if instruction is re-thought based on competencies, and if new materials are thoughtfully integrated, it is possible to design rich and engaging instruction that goes beyond the possibilities of traditional approaches.

This chapter explored several considerations related to the creation of competency-based education, including how competencies can be used to inform training, how they can be combined to create a curriculum and the challenge of how different people may interpret competencies differently (requiring "hermeneutic" skills to interpret texts appropriately). In addition, the chapter introduced learning design concepts relevant to CBE. It was explained that training strategies should be employed that are suited to types of learning implicit in competencies and some different frameworks for identifying these types were presented. The identification of types of learning facilitates effective selection of training strategies. Differences between "generative" and "supplantive" strategies were discussed and it was indicated that generative approaches are appropriate for more experienced learners but tend to increase cognitive load. Supplantive approaches reduce cognitive load and are suited for learners who may be very unfamiliar with the content of training.

Templates for constructing training programs were presented, including "events of instruction" models and the whole-task model.

As e-learning is increasingly expected to play a large role within future training programs, the characteristics of this training delivery method were discussed in detail. Although e-learning can be used to enhance the quality of aviation training, the unfortunate reality is that most decisions regarding the implementation of e-learning are based on cost savings. Nowadays it is common for classroom training to be eliminated completely in favor of e-learning, and all too often the e-learning courses are rapidly repurposed, poor-quality versions of classroom courses, but without the component that is often the most powerful part of classroom teaching—the human instructor.

Pause for a moment and reflect on all of the educational experiences you have had in your life. Which learning experiences have had a lasting and memorable impact? It is probably safe to assume that the most powerful experiences were related to a skilled human instructor and related to the application of learning to real-world challenges—and did not take place while sitting in front of a computer. This is not to say that e-learning is ineffective or useless; the point is that e-learning is a powerful resource that a skilled instructor can use to complement and enhance a training program. E-learning that entirely replaces classroom instruction may have negative impacts on training outcomes.

Within competency-based education, e-learning can present pre-training before learners arrive at the training center and can also be applied to desktop simulators, homework and digital resources that will be referenced during training. After learners have left the training center, e-learning can be used to distribute snap-courses to keep skills fresh throughout the year. This approach should allow e-learning to enrich the training curriculum, taking advantage of e-learning strengths and avoiding the quality pitfalls that all too often are associated with the integration of e-learning.

The chapter finished with a brief overview of the ICAO instructor competency framework. The four competencies cover concepts that have been treated in this chapter, but also draw attention to the importance of preparation for effective training and administrative tasks. It was pointed out that in some implementations of CBE, trainers find that their administrative duties increase.

Chapter 9

Assessment

Questions Considered in this Chapter:

- How is assessment conducted within a competency-based education system? How is this different from assessment within traditional approaches?
- What evidence is assessment based upon? How can training managers ensure that the evidence is directly related to a competency?
- What can be done to improve the reliability of competency-based assessment?

Society's confidence in the competence of air traffic controllers, cabin crew, aircraft maintenance engineers and pilots is crucial for the aviation industry. Whether to ensure that new entrants possess base-level skills or to confirm the competence or development of existing staff, assessment is a key to promoting integrity and confidence in the industry. In this chapter we consider assessment within a competency-based education (CBE) system. CBE assessment is characterized by systematic reference to competency statements/texts ("competencies"). Competency-based assessment is a form of "criterion-referenced" assessment, with criteria drawn from the content of competencies. Assessment of a person's competence, as specified in relevant competencies, must be based on evidence. We consider kinds of evidence appropriate to competency-based assessment, including evidence relating to performance and evidence relating to knowledge and understanding. The extent to which evidence can be regarded as evidence of a competency raises the problem of validity. Assessment research and theory have identified numerous aspects of validity that should be taken into account by assessment designers and assessors. The chapter addresses the practical problem of reliability and ways to improve it. The relationship between diagnostic, formative and summative assessment will be examined, and finally the process of conducting assessment will be considered.

Before turning to the concept of competency-based assessment as such, four preliminary issues will be discussed briefly. We will examine the relationship between evaluation and assessment before looking at the differences between criterion-referenced and norm-referenced kinds of assessment. The problem of assessing levels of performance will be considered, followed by the relationship of training and prior experience to competency-based assessment.

Evaluation and Assessment

In this book we adopt the convention of differentiating evaluation and assessment. Evaluation is, as the term suggests, the process of determining the value of something. Stufflebeam and Shinkfield (2007) explained that earlier definitions of evaluation focused on whether or not objectives of a program were achieved. They argued, however, that evaluation should be concerned with not only outcomes as specified in documents like competencies, but also the extent to which such official statements of objectives are themselves adequate to the broader goals of a program. For Stufflebeam and Shinkfield, evaluation is concerned with whole programs and their contexts, as reflected in their definition: "evaluation is the systematic process of delineating, obtaining, reporting, and applying descriptive and judgmental information about some object's merit, worth, probity, feasibility, safety, significance, and/or equity" (16). This definition of evaluation extends beyond the question of whether individuals can be deemed competent or not against criteria contained in competencies. We accept that the term *evaluation* is better suited to program-level measurement and judgment, which, in the context of CBE, would include questions about the effectiveness of program design, and potentially the effectiveness of CBE as an approach that responds to the aviation industry's need for an effective platform for training and assessment. It goes beyond the scope of this book to examine program evaluation, but it is obviously an important process that can be applied at many levels beyond that of individual achievement.

Our use of the term *assessment* is closer to earlier definitions of evaluation as the measurement of learner achievement of program objectives. In our case, assessment has an individual focus and concerns the extent to which individuals can be deemed competent when measured against the description of tasks contained in competencies.

Criterion-Referenced versus Norm-Referenced Assessment

Chapter 1 explained the relationship between criterion-referenced assessment and CBE. Some forms of assessment, known as *norm-referenced assessment*, are designed to determine the standing of an assessment candidate relative to normal performance of a reference group. For example, assessment may be used to identify top, bottom and average performers within a group, perhaps according to a normal distribution curve, and assessment results are reported in terms of an individual candidate's performance against the group norm. Traditional high-stakes examinations of final-year school students are norm-referenced. Every year, a fixed proportion of students will fall into each band. Note that such assessment is not *directly* related to the extent to which students demonstrate performance or understanding against subject-based criteria. The potential is there for a whole group of students to demonstrate very high levels of understanding, yet a fixed proportion will be relegated to lower bands and a fixed proportion to the higher. This kind of assessment determines one's standing relative to a group's performance.

In contrast, *criterion-referenced assessment* seeks to judge a candidate's performance in relation to pre-specified criteria (Glaser 1963). In this kind of assessment, it is irrelevant how many candidates in a cohort meet the criteria or not. It is possible that all or no candidates may be assessed as achieving the criteria. This kind of assessment cannot inform users of assessment reports how the candidate stands with respect to other candidates. There is no information about the "norm" of the group, and therefore no way to measure the individual candidate's standing with respect to it. Rather, this kind of assessment measures and reports on the assessee's level of performance against criteria drawn from the domain of learning. Assessment in CBE is a form of criterion-referenced assessment, in which the domain of learning is an occupation. The criteria describe occupational skills and knowledge targeted for learning. In CBE, competency documents supply the specifications of criteria that serve as the reference point for assessment design.

Minimum-Competency versus Graded Assessment

An issue that plagues some implementations of CBE is whether it is useful and/or justifiable to differentiate levels of competence. That is, can performances be graded in some way with respect to pre-specified criteria? Chapter 1 introduced the concept of minimum-competency assessment (Jaeger and Tittle 1980). This is the idea that it is desirable to focus on whether or not a person can be judged as competent or not, rather than attempting to determine how far above or below competent they are. The principles of CBE support the concept of reporting assessment judgment in binary terms of competent or not competent, since the information contained in competencies generally describes a single state of competence, without the kind of additional elaboration that would be necessary to guide assessment designers and assessors to differentiate performances above and/or below "competent." In some implementations of CBE, the affinity between the principles of CBE and the concept of minimum-competency assessment is foregrounded. However, this practice does not necessarily meet the needs of all system stakeholders. Recruiters may be interested in determining just how competent a potential employee is after initial training, while human resources departments and leadership may be interested in how well an employee performs in continuing training and development. Assessees themselves may prefer to be recognized for exceptional performance. Hence, minimum-competency assessment may not serve all stakeholders, although conceptually it is consistent with other aspects of CBE.

The principles of criterion-referenced assessment were developed independently of CBE, and from the start this kind of assessment was amenable to grading. For example, Glaser (1963) explained that:

> Underlying the concept of achievement measurement is the notion of a continuum of knowledge acquisition ranging from no proficiency at all to perfect performance. An individual's achievement level falls at some point on this continuum as indicated by the behaviors he displays during testing.

> The degree to which his achievement resembles desired performance at any specified level is assessed by criterion-referenced measures of achievement or proficiency. (519)

So although the content of competencies may present competence as something that will be either present or not in the assessee, the practice of grading is consistent with the principles of criterion-referenced assessment. Additional information will always be required, however, to allow assessment designers and assessors to make differentiated judgments of competence. Glaser (1963) indicated that specific behaviors along his continuum of competence can be identified and specified to allow differentiated assessment of performance.

Assessment, Training and Prior Experience

One of the characteristics of the competency-based approach is that the specification of competence in terms of outcomes is separated from questions about how competence is developed. This means a judgment of a person's competence is unrelated to the kinds of learning experiences they underwent on the way to becoming competent. An implication of such a separation is that training providers have great freedom about program design and training strategies. Another is that formal training need not be a route to achievement of competence. Any of a range of experiences can serve to develop competence, from on-the-job to formal programs of learning (McMullan et al. 2003). Being a graduate of this or that kind of training or having certain job experiences should not be a factor in the design and processes of assessment. The design and conduct of assessment in CBE must reckon with the possibility of multiple routes to competence, and evidence of competence must not be identified with participation in particular experiences. Methods for assessment of prior experience and learning are relatively sophisticated, providing assessment designers and assessors scope to undertake this kind of assessment with confidence. The portfolio methodology, for example, is well understood and an appropriate approach to gathering evidence for competency-based assessment (McMullan et al. 2003).

Evidence for Assessment

Assessment against criteria demands some kind of evidence that the assessee can perform or has knowledge as specified. Rutherford's (1995) discussion of sources of evidence for competency-based assessment distinguishes between evidence of performance and evidence of knowledge or understanding. This distinction has shown up in different ways in this book. In Chapter 1, differences between the "competence" and "performance" traditions of CBE were discussed. The competence tradition focused on internal attributes that manifest in competent work, and the performance tradition focused on observable behavior. The distinction also resonates with the differences between behavioral approaches and cognitive approaches to learning, as well as between "skill" and "knowledge." Rutherford identified eight types of evidence in the "performance" category. The first is *direct observation* of the candidate in an actual work environment. Evidence from direct observation may be regarded as the "holy grail" of competency-based assessment. According to Rutherford:

> Because CBA [competency-based assessment] aims to assess and measure job related outcomes, the best form of evidence is a demonstration of what naturally occurs in the workplace. This is called natural evidence and is the most accurate when judging whether or not a person can actually perform the overall function. (164)

The strong focus of CBE on knowledge and skills in application accounts for the value placed on evidence based on observation of people doing their work. Competency texts generally offer some description of the task in behavioral terms, in principle allowing an assessor to compare actual performance with the behaviors described in the competency. The assessor can read from the competency a description of competence (e.g., elements) and how well the assessee is supposed to perform (e.g., performance criteria). A good example of direct observation evidence is training cabin crew on passenger evacuation from an aircraft. During training of evacuation procedures, cabin crew undertake a number of training sessions aimed at both knowledge and skills. These sessions include (a) reasons for evacuation, such as an uncontrollable fire on the aircraft, or evacuation *commanded* by the captain; (b) reasons for using or blocking a particular exit (blocking an exit may be necessary during a ditching, where many large high-capacity aircraft are designed to float tail low in the water—remember the Hudson River A320 evacuation—requiring rear exits *not* to be used to reduce flooding of the aircraft); (c) commands to be used during the evacuation, so passengers are given clear instructions; and (d) actual physical operations of all emergency exits, including doors, over-wing exits and even cockpit windows. During final assessments, cabin crew trainees will be required to correctly respond to external or internal signals to evacuate an aircraft, culminating in a full aircraft evacuation. During this process, assessors will observe each individual's actions and decisions during the evacuation.

A widely recognized limitation of direct observation as a form of evidence is that in practice, observations can only capture a brief slice of the assessee's behavior. For many reasons, it is possible for an assessee to "fluke" a competent performance, or lapse into less competent practices over time or when they think they are no longer under observation. For this reason, advocates of direct observation advise collecting this kind of evidence at intervals and/or complementing observation with other kinds of evidence.

Examination of the *outcome of an assessee's work or other types of output* is another source of evidence of performance. Rutherford (1995) explained that:

> The outcome or end product of a candidate's work is an ideal way of measuring his or her effectiveness at job-related skills. Such evidence might be a product, a document, or a service performed for or to clients, and rather than assess the process [i.e., through direct observation], an assessment is made of what the processes are designed to achieve. (165)

An example of this kind of evidence was that collected in performance-based teacher education (PBTE, an early version of CBE discussed in Chapter 1). In PBTE, teacher competence was in part judged on the basis of the assessment results of the students they taught. In this example, it is a complex "product" of the teachers' competence that serves as evidence. For instance, during the early training of aircraft maintenance engineers, some training organizations require students to build a toolbox. When the toolbox is finished, it is the actual tolerances of workmanship such as angles, rivets, filing and final measurements that are the measure of final performance.

Rutherford (1995) presented work-related *simulations* as an appropriate source of evidence "when it is impractical for reasons of security, safety or opportunity to have the candidate demonstrate competency in a realistic work setting" (166). This kind of evidence of competence is highly appropriate to occupations within the aviation industry, because of the inherent risks and costs of failed assessee performance. One of the best examples in aviation is in the area of pilot training. For decades, the aviation industry has realized that while training for scenarios in real aircraft may well prepare a pilot to deal with specific problems, the evidence also shows that some scenarios or maneuvers kill more pilots than real-life events. Given the lessons learned over many years, the aviation industry has moved toward simulation training of pilots as the fundamental instruction process. In most airlines, pilots do not ever practice any malfunctions or maneuvers in real aircraft! However, it has also been recognized that some of the smaller turboprop aircraft often do not have the availability of simulators that the larger high-capacity jet aircraft do. This limitation places severe restrictions on the training of pilots. For instance, the "examination of a serious incident involving a 30-seat passenger aircraft found that around 13 malfunctions could be practised in the real aircraft with safety. This is in contrast to the 225 malfunctions that could be practised in the simulator" (Mavin and Murray 2010, 282). According to Rutherford (1995), using simulation as a source of evidence of competency should be complemented with other kinds of evidence, including actual work performance, once the risks associated with some kinds of work are minimized through assessees' demonstration of competence in a simulated environment.

Further removed from the actual work setting are *tests of skills, projects* and *assignments* that may serve as sources of evidence of competence. Tests can allow detailed inspection of an assessee's performance that may not be possible in actual work situations. Projects and assignments can generate products that can be assessed. One of the classic uses of knowledge-based exams is in pilot and cabin crew training during annual emergency procedure (often referred to as EPs) training and assessment. Here pilot and cabin crew, who generally train together, will undertake a quiz to test the understanding of aircraft systems prior to simulator detail. These exams test knowledge of the number and positioning of emergency equipment such as fire extinguishers and personal breathing equipment, as well as knowledge of limitations in the use of individual equipment. For example, certain fire extinguishers cannot be used on electrical fires. As cabin crew are required to have specific medical training, cabin crew exams will ask specific medical and first aid questions. These three forms of evidence belong to a group of "indirect" sources of evidence that, in the CBE context, would need to be complemented with evidence from other sources, including direct evidence sources, to build a complete picture of an assessee's competence.

The other category of evidence described by Rutherford (1995) relates to knowledge or understanding. The need to collect evidence of knowledge confronts the assessment designer and assessor with a problem that has surfaced a few times in this book: how do we form a picture of underlying processes that cannot be directly observed? Rutherford suggested that the application of such knowledge is the opportunity to take a measure of its nature and extent. Methods to allow observation of application include the following:

- work-related assignments
- projects
- tasks
- role plays
- case studies
- questioning
- tests
- documents.

Work-related *assignments, projects and tasks* form a group of opportunities to observe knowledge in application. These methods base the activity in actual work conditions, but specifically allow the application of knowledge to be examined through written reports by the assessee. *Role plays and case studies* also offer opportunities to observe the application of knowledge. Assessment design must identify the kind of knowledge required (supplied by analysis of the learning implicit in the task) and select or build the role play or case study to provide ample scope for exercise of the identified knowledge.

Work-related *questioning* allows assessment of application of knowledge to a wider range of settings than that afforded by immediate work conditions. A related method is the use of *tests* requiring written reports such as essays, short-answer questions and multiple-choice questions. These more traditional methods of gathering assessment evidence do have a place in contemporary training where complex knowledge is being measured. As a technique used in combination with other methods, written tests help to fill out the picture of the assessee's understanding of complex situations. A final method involves evidence sourced from documentation associated with previous work-related achievement. Rutherford (1995) conceded a number of shortcomings with this form of evidence, but such documents may happen to be available and therefore contribute to an assessment of complex knowledge and understanding.

A number of principles or "rules" of evidence have been identified to guide the design and conduct of competency-based assessment. *Sufficiency of evidence* refers to having access to enough evidence on which to confidently make a judgment of competency. In general, multiple sources of evidence and evidence collected over a period of time produce sufficient evidence. *Authenticity* refers to the extent to which the assessor can be sure that the evidence was really produced by the assessee. Evidence of performance as described by Rutherford (1995) allows authenticity to be demonstrated, while other sources of evidence such as assignments, online assessment and portfolios may raise questions of authenticity. *Currency* is a third rule of evidence which, as the name suggests, refers to how up to date the evidence on hand actually is. A fourth rule of evidence, *validity*, is a major consideration and will be treated next.

Validity of Assessment

A perennial concern for assessment designers and assessors is the extent to which evidence can be regarded as *evidence of* true competence as described in a competency text. Validity refers to the fit between evidence and the subject of the evidence. High validity implies that the assessment being used is measuring what it claims to measure. Low validity would be applied to an instrument that measures something that is not related to the actual need of the assessment task (Linn and Gronlund 1995). Even though there are many types of validity, four main types are used in education: content validity, criterion-related validity, construct validity and face validity.

Content validity relates to the relationship between the assessment and instructional outcomes (Linn and Gronlund 1995). If there is a high level of correlation between what is being taught and what is being assessed, it is said to have high content validity. Low content validity can be a reason behind student frustration ("the test was nothing like what we were taught"). However, high content validity does not necessarily mean that the assessment is sound.

Criterion validity relates to how well scores from a particular assessment can relate to the theoretical concepts of performance. Criterion validity is often viewed as being made up of (a) predictive validity and (b) concurrent validity (Peng and Mueller 2003b). Predictive validity, as the name implies, is a measure of how well an assessment can predict future performance. For example, in many airline pilot assessment tasks, there are numerous fields to assess, including knowledge, communication and management. It might be discovered that while knowledge and communication are important for being a captain, the most likely predictor for a first officer passing a simulator assessment is their management ability.

Concurrent validity is an indication of how well a new assessment measures up against an older, but validated assessment. That is to say, "concurrent evidence of criterion-related validity indicates the test's degree of association with a current criterion. Concurrent validity differs from predictive validity only in that the criterion is concurrent" (Peng and Mueller 2003b, 214). For example, say assessment A had been used for many years and had been tested and determined valid. If a new assessment B is introduced, one way to measure its validity is to compare its results with those of the older assessment A. That is, both assessment A and B are conducted "concurrently" and compared.

Construct validity relates to how well an assessment measures a particular construct (Peng and Mueller 2003a). That is, to what level does the assessment allow the assessor to make inferences from what he or she has observed? For example, in NOTECHS (see Chapter 4), there is a category called "Leadership and managerial skills." The elements of this category are (a) Use of authority and assertiveness, (b) Providing and maintaining standards, (c) Planning and coordination, and finally (d) Workload management. If measuring these four units shows high levels of correlation between the units and the category, this would have high construct validity. It is interesting to note that over half a century ago, Loevinger (1957) proposed that construct validity contained both content and criterion validity. Thus, "construct validity is at the heart of the validation process—an evidence-gathering and inference-making progression" (Peng and Mueller 2003a, 183).

The final form of validity is *face validity*. It is a test to see how a particular assessment is viewed from an individual perspective, rather than using psychological testing. Even though face validity is often used in aviation, it has not been without its controversy:

> ... illegitimate usage has cast sufficient opprobrium on the term as to justify completely the recommendation that it be purged from the test technicians' vocabulary, even for its legitimate usage. The concept is the more dangerous because it is glib and comforting to those whose lack of time, resources, or competence prevent them from demonstrating validity (or invalidity) by any other method. Moreover, it is readily acceptable to the ordinary users of tests and its acceptance in these quarters lends the concept strength. This notion is also gratifying to the ego of the unwary test constructor. It implies that his knowledge and skill in the area of test construction are so great that he can unerringly design a test with the desired degree of effectiveness in predicting job success or in evaluating defined personality characteristics, and that he can do this so accurately that any further empirical verification is unnecessary. So strong is this ego complex that if statistical verification is sought and found lacking, the data represent something to be explained away by appeal to sampling errors or other convenient rationalization, rather than by scientific evidence which must be admitted into full consideration. (Mosier 1947, 194)

Over the years the use of the term *face validity*—while still not well received by many academics—is gaining at least some level of acceptance among assessment researchers:

> Although some researchers argue that a high degree of face validity is desirable, whereas others contend that it is undesirable, researchers in both camps agree that face validity functions as a moderator variable that may influence the concurrent and predictive validity of a test. (Bornstein 1996, 983)

Reliability of Assessment

Another issue for assessment in CBE is the extent to which assessment decisions are consistent between different learners at different times and in different settings undertaken by different assessors. The term *reliability* focuses not on how a test can effectively measure performance, but rather how the assessment scores change, or (preferably) don't change, over assessment sessions, typically with another person carrying out the assessment on individuals performing the same tasks. It is focused on "the consistency of assessment results" (Linn and Gronlund 1995, 48).

Baker and Dismukes (1999) identified three methods that could be used to improve inter-rater reliability: rater-error training, performance dimension training, behavioral-observation training and frame-of-reference training. *Rater-error training* involves explaining to the assessor different errors that can possibly be made during assessment, such as halo effect, horn effect, central tendency, leniency, severity, primacy and recency (Baker and Dismukes 1999; Woehr and Huffcutt 1994). *Horn* and *halo* effect occur when final performance assessment is affected by evidence of poor or good performance that is small compared with the entire assessment—for example, an overall good performance being marked down heavily because of one mistake. *Leniency* and *severity* can occur if the assessor provides disproportionately high or low marks overall. Pilots have been heard to make comments such as, "I like being checked by Fred, he is an easy marker," and sometimes training managers will have an issue with pilots calling in sick on simulator assessments because of a check captain who has a reputation of being a strict marker. In contrast, *central tendency* occurs when the assessor tends to use the middle grades; that is, they rarely fail anyone or provide exemplary marks. Finally, *primacy* and *recency* arise when the assessor is affected by how an individual begins or finishes an assessment (Baker and Dismukes 1999).

Performance dimension training, the second method of inter-rater reliability training, involves familiarizing the assessor with the assessment instrument that they will be using (Baker and Dismukes 1999; Woehr and Huffcutt 1994). Another way of improving inter-rater reliability is with *behavioral-observation training*. Here the assessor is trained in differentiating, which focuses on examination of behavior, rather than evaluation. That is, what behavior is related to management or communication? Woehr and Huffcutt (1994) observed that the "judgment processes include the categorization, integration and evaluation of information, while observation processes include the detection, perception and recall or recognition of specific behavioral events" (192).

A final method that has been used in the past in an attempt to improve inter-rater reliability is *frame-of-reference training*. For example, pilots or cabin crew might watch a video and practice assessing the behavior shown in the video. On completion, the trainer will attempt to obtain a group consensus and "moderate" the grades provided; for example, the trainer may facilitate a discussion on whether a grade is too high or too low. Frame-of-reference training provides the "multidimensionality of performance, defining performance dimensions [and] provides a sample of behavioral incidents representing each dimension" while also allowing "practice and feedback" (Woehr and Huffcutt 1994, 192). It has been proposed that frame-of-reference instruction is the most effective form of inter-rater reliability training (Baker and Dismukes 1999; Woehr and Huffcutt 1994).

The ability to accurately assess performance is foundational to the aviation industry. In an attempt to objectify the assessment process, many of the current assessment approaches break performance into small measurable components; this approach is often referred to as reductionist. It is generally thought that this approach to assessment improves both validity and reliability of assessment. Yet, if at the end of an assessment the measures cannot be used to do the evaluation—that is, make judgments about individual or team performance, or the effectiveness of the training—the process is not working as was initially planned. For example:

> when we review the previous cyclic program [i.e., the assessment conducted six months earlier] to brief for the next one, there is no way that I can go into the data of our current competency system and tell the check captains where the problems lie in the system. I have anecdotal evidence, I have a couple comments, *but I cannot put my hand on my heart and say that the data is giving me this. It's just useless.* [emphasis added] (PIL5)[1]

Here the chief pilot of an airline that currently uses detailed competencies is concerned that their current assessment program, built on 50 pages of competencies, does not enable him to make judgments about future training programs or other pilots.

One of the proposed issues with current performance assessment is the theory of performance assessment itself. According to Goevarts et al. (2007), raters who do not have clear and common understanding of the theoretical nature of performance "are unlikely to hold common definitions of work performance. Evidence from the field of performance appraisal has shown that framing performance assessment as a psychometric problem may lead to preoccupation with raters' errors" (249). Intrinsic in the issues that Goevarts et al. outline is that:

---

1 PIL5: Training Manager, Southeast Asia.

> Assessment instruments often reflect designers' implicit theories about performance and describe performance dimensions and standards against which observed behaviour is to be measured. Although these external guidelines may be useful, they are generally unable to account for raters' judgments and decisions. (Goevarts et al. 2007, 249)

Other studies of the reliability of assessment—primarily pilot assessment—have shown that large variations occur in performance assessment; that is, inter-rater reliability is poor (Brannick, Prince and Salas 2002; Mavin, Roth and Dekker 2013; Smith, Niemczyk and McCurry 2008). According to one study, even after assessors had received years of training, only small improvements in inter-rater reliability are measured (Holt, Hansberger and Boehm-Davis 2002).

One of the key issues that still has not been resolved is in relation to the measures used to assess inter-rater reliability. Many techniques used to measure inter-rater reliability assume that variations can, in many cases, be explained through errors made during assessment. However, a recent study into pilots assessing the performance of other pilots showed that the supposed "errors" that pilots made during the assessment process were actually well reasoned.

Summative, Formative and Diagnostic Assessment

Assessment is usually thought of as something that happens after training—taking a measure of the learning that the assessee can demonstrate. This kind of assessment is called "summative" assessment. It confirms learning and underpins the awarding of licenses and qualifications. But assessment processes can serve a range of purposes apart from confirmation and credentialing. Bloom, Hastings and Madaus (1971) made the influential case that assessment can serve "formative" and "diagnostic" purposes as well. Learning can be promoted when learners can gauge their own progress and assessment during the learning process can provide this "formative" feedback. Assessment can also give trainers and learners a baseline measure of learners' prior knowledge; such "diagnostic" assessment can be used to develop formative and summative assessment later (Bloom et al. 1971). In this section we consider these three forms of assessment, beginning at the pre-training phase with diagnostic assessment then moving to formative and finally summative assessment.

According to Bloom et al. (1971), "Diagnostic evaluation performed prior to instruction has placement as its primary function; that is, it attempts to focus instruction by locating the proper starting point" (87). In terms of diagnostic types of assessment, Smith and Ragan (2005) distinguished between entry skills assessment and pre-assessments. Entry skills assessment refers to measurement of prerequisite knowledge. Where instruction in particular competencies assumes the prior possession of other competencies, it may be appropriate to confirm these prerequisite competencies in a formal way. Smith and Ragan made the point that even if learners have graduated from training programs designed to develop prerequisite competencies, these learners may not have had an opportunity to practice their learning or may simply not have retained the learning. To the extent that the present training assumes the prior acquisition of specific competencies, particularly ones for which training and assessment occurred some time in the past, it makes sense to confirm the current competency of learners in a new program. *Entry skill assessment* is the term given to this confirmation of possession of identified prerequisite competencies.

Pre-assessment refers to diagnostic assessment that focuses on what learners already know about the new topics (Smith and Ragan 2005). This kind of assessment has the very practical goal of allowing the trainer to adjust the training program to take prior knowledge into consideration. Trainers can devote more time to areas relatively unknown to learners and reduce the amount of time on material that learners already know. This kind of diagnostic assessment may also serve to heighten learners' attention to the new content, priming learners for instruction. It is also possible that pre-assessment will reveal that certain learners already possess the target competency.

Formative assessment—also called "assessment for learning"—is used to improve learning, but can also help to improve other aspects of training such as training strategy and program design (Bloom et al. 1971). Of course, trainers and teachers have always undertaken formative assessment in the normal course of their interactions with learners, in the form of feedback. Research into formative assessment offers a systematic approach to formative assessment. Drawing from a large-scale study of formative assessment research and practices in Britain, Black and Wiliam (2012) identified four practices for realizing the potential of formative assessment to boost learning:

- dialogue
- feedback through marking
- peer- and self-assessment
- formative use of summative tests.

Dialogue, or questioning, is a traditional formative assessment practice. Here, learners are asked questions about their understanding of a topic and their answers generate reflection and discussion that guide further learning. Black and Wiliam (2012) reported that this technique is enhanced by allowing extra time for learners to make responses and by framing questions so that they elicit rich responses for discussion. *Feedback through marking* refers to extensive use of comments in assignments. Black and Wiliam stressed that effective comments include explicit guidance for improvement. *Peer- and self-assessment* is formative assessment by learners. Self-assessment gives learners responsibility for their own learning. This kind of assessment is effective when learners have access to, and understanding of, the criteria for learning. In CBE, this means learners are exposed to competencies and can interpret them at least enough to gauge their own progress against criteria expressed in the documents. In peer-assessment, learners adopt the roles of trainer and assessor with respect to other learners. In this process, learners develop deeper understanding of criteria as they compare the work of other learners with their own.

Finally, with regard to *formative use of summative tests*, Black and Wiliam (2012) identified three ways to derive formative assessment from summative testing processes. First, learners are given a list of knowledge and/or skills that may be tested summatively and are asked to identify areas in which they consider themselves to be stronger or weaker. This approach encourages learners to create strategies for managing their preparation for summative assessments. Second, peer-assessment of learner responses to summative assessment gives learners an opportunity to assess each other's summative work prior to official results being made available. In this practice, learners can be encouraged to develop the criteria for assessing the material, and peers have an opportunity to defend their actual responses with each other. Third, asking learners to write assessment items themselves and provide answers has been found to improve overall assessment performance.

Summative assessment generates information about achievement for learners, trainers and stakeholders outside the learning situation. In CBE, such assessment confirms that learners are competent as specified in the target competency or competencies. Summative assessment is effective to the extent that sources of evidence are appropriate; that valid assessment has been constructed to yield the evidence; and that assessment instruments, processes and assessor professional development are such that reliable assessment judgments may be expected from the process.

Note that although this section has treated diagnostic, formative and summative assessment as having a fixed relationship with each other, the reality of assessment in a complex industry such as aviation means that the "learners" we have been referring to are always moving into or out of learning and assessment processes as they become competent, need to maintain their competence or branch out into new subfields or levels of work. This summative assessment can serve as diagnostic and formative assessment material for new phases of learning. The three basic forms of assessment feed into each other in the context of the careers of aviation professionals.

Conducting Assessment

The process of conducting assessment calls for input from assessment designers—who determine evidence requirements against competencies, address validity concerns and promote reliability—and implementation by assessors. Assessors are usually experts in the occupational area under consideration and, in addition, possess competence in the implementation of assessment. Assessor expertise is an indispensable factor in an effective assessment system, because there is always an element of judgment in arriving at an assessment decision.

Regardless of the comprehensiveness of the competency text and the expertise brought to bear by assessment designers, evidence must always be weighed up, especially in complex performances typical of the aviation industry. That is, no matter how exhaustive the design of the assessment itself, an expert judgment that takes many factors into account is still required. Effective assessment design will produce guides that assist individual assessors in arriving at particular assessment decisions (and to boost reliability). Assessment decisions may be formalized based on input of multiple assessors, but component decisions remain a matter of judgment by individual assessors.

The aviation industry has produced a statement of the competencies required to carry out competency-based assessment. According to the ICAO (2006), three competencies articulate the work of the assessor or "examiner": "Gather evidence," "Evaluate evidence" and "Report assessment decision." The elements for these competencies have been specified as follows:

1. Gather evidence.
    1.1.   Establish a working relationship with the candidate.
    1.2.   Interpret competency standards.
    1.3.   Apply assessment techniques and tools.
2. Evaluate evidence.
    2.1.   Ensure validity of evidence gathered.
    2.2.   Ensure reliability of evidence gathered.
    2.3.   Establish assessment decision.
    2.4.   Provide constructive feedback to the candidate
3. Report assessment decision.
    3.1.   Record assessment results.
    3.2.   Provide candidate with future training plan, if applicable.
    3.3.   Review assessment process to improve validity and reliability.
    3.4.   Process relevant documentation. (ICAO 2006, 5-A-3)

These competencies draw attention to a number of practical issues facing assessors. First, effective conduct of assessment requires a "working relationship" with the assessee. It is desirable for the assessee to feel comfortable with the assessor and the process, so the assessee can focus on the assessment and not be overly distracted by the presence of the assessor or the occasion of assessment. Although assessment designers produce resources to guide and document assessment based on their interpretation of relevant competencies, the assessor must also possess understanding of the same competencies. In theory, since competencies aim to articulate the occupational area in which the assessor has expertise, the assessor should be able to read the relevant competencies without difficulty. However, some research suggests that this assumption may be flawed (Hodge 2014). The third element in the "Gather evidence" competency is "Apply assessment techniques and tools," which refers to using assessment materials designed for the guidance and documentation of assessment.

The second competency, "Evaluate evidence," links with concepts covered in this chapter: evidence, validity and reliability. This competency gives assessors responsibility to contribute to validity and reliability of assessment, a goal that assessment designers also aim for when they create assessment items, guidelines and documentation for use in assessment. "Establish assessment decision" refers to that key exercise of assessor expertise mentioned above. This is the most crucial and difficult part of the assessor's work, a role only the assessor can accomplish and one which is supported only up to a point by the work of assessment designers. Finally, the assessor is expected to provide feedback on the performance of the assessee. In most implementations of CBE, unsuccessful assessees are invited to be reassessed, and it is the feedback provided on an unsuccessful attempt that gives the assessee the formative information required to be successful on the next attempt.

The final phase in the conduct of assessment is captured in the third competency, "Report assessment decision." In summative assessment, reporting is an important part of the process because of the interests of external stakeholders in these assessment decisions. Templates will have been designed for use in these assessments; the templates record in more or less detail the components of the assessor's decision against aspects of the competency or competencies. The reporting process is crucial for credentialing purposes. The report will be processed and stored, serving as a basis for issuing certificates and licenses and for monitoring employee progress and achievement.

Conclusion

This chapter has introduced key dimensions of competency-based assessment to guide the design and conduct of assessment. The importance of assessment to the aviation industry cannot be stressed enough. CBE presents a range of challenges and opportunities for improving assessment in the industry. A few clarifying points were made. The decision to distinguish assessment of individual performance from evaluation of broader factors (such as program design and the fitness of CBE for the industry) was explained. The difference, too, between criterion-referenced and norm-referenced assessment was highlighted. Assessment in CBE is criterion-referenced, with the criteria being drawn from the competencies. The topic of graded versus minimum competency assessment was discussed. Criterion-referenced assessment need not be of the binary (competent versus not competent) variety. However, if the assessment is related to predetermined criteria, a large amount of additional descriptive work is required to identify these criteria. We discussed the flexibility inherent in competency-based assessment in relation to the kinds of learning that assessees have experienced. Assessment in CBE in principle does not concern itself with the specific learning experiences that lead to competent performance, suggesting that assessment designers and assessors should be amenable to innovative ways of demonstrating competence, such as portfolios.

Assessment in CBE emphasizes sources of evidence that are close to actual workplace conditions. Direct observation of assessees doing the work in which they are being assessed is the ideal, but there are numerous limitations to take into account, including the fact that assessment in work conditions can only ever take a sample of performances. We also identified sources of evidence of knowledge or understanding. The knowledge underpinning performance is especially important to measure in highly sophisticated environments such as those found throughout the aviation industry. Validity is a key "rule" of valuable evidence. The evidence must be evidence of the competency under consideration. The fit between the specified performance and acceptable evidence of that performance raises a number of questions, and a range of forms of validity have been distinguished to help answer these questions. The reliability of assessment decisions is a practical problem for all assessment. To what extent will different assessees, assessed by different assessors, at different times and under different conditions, be assessed in a consistent way? This important issue was discussed and measures to strengthen reliability identified.

Assessment is not only about a final measure of achievement. It can be used to diagnose current learning prior to entry into training programs (to confirm possession of prerequisite competencies and to alert trainers to prior knowledge relating to new topics) and also in a formative way to systematically provide a learner with feedback and other guidance. Summative assessment is a third mode of assessment that occurs at the end of training to measure achievement and/or for credentialing. The conduct of assessment was discussed. It was pointed out that the expertise of assessors is indispensable for making a judgment of competence, because all evidence and support mechanisms are necessarily limited, and the assessor must be personally confident in his or her assessment of an assessee's performance. The ICAO (2006) competencies for examiners were discussed, suggesting ways they align with key assessment concepts treated in the chapter.

Part III

References

Anderson, L.W. and D.W. Krathwohl, eds. 2000. *A Taxonomy for Learning, Teaching and Assessing: A Revision of Bloom's Taxonomy of Educational Objectives*. 2nd edn. New York: Pearson.

Annett, J. and N.A. Stanton. 2000. *Task Analysis*. London: Taylor & Francis.

Baker, D.P. and R.K. Dismukes. 1999. Training pilot instructors to assess CRM: The utility of frame-of-reference (FOR) training. In *Proceedings of the 10th International Symposium on Aviation Psychology*, ed. R. Jensen, 291–300. Columbus: Ohio State University.

Bernard, R.M., P.C. Abrami, Y. Lou, E. Borokhovski, A. Wade, L. Wozney et al. 2004. How does distance education compare with classroom instruction? A meta-analysis of the empirical literature. *Review of Educational Research* 74(3): 379–439.

Black, P. and D. Wiliam. 2012. Assessment for learning in the classroom. In *Assessment and Learning*, ed. J. Gardner, 11–32. 2nd edn. London: Sage.

Bloom, B.J., M.D. Englehart, E.J. Furst, W.H. Hill and D.R. Krathwohl. 1956. *Taxonomy of Educational Objectives*. Handbook I, *Cognitive Domain*. New York: McKay.

Bloom, B.S., J.T. Hastings and G.F. Madaus. 1971. *Handbook on Formative and Summative Evaluation of Student Learning*. New York: McGraw-Hill.

Bornstein, R.F. 1996. Face validity in psychological assessment: Implications for a unified model of validity. *American Psychologist* 51(9): 983–4.

Brannick, M.T., C. Prince and E. Salas. 2002. The reliability of instructor evaluations of crew performance: Good news and not so good news. *International Journal of Aviation Psychology* 12(3): 241–61.

Cavanaugh, C.S. 2001. The effectiveness of interactive distance education technologies in K-12 learning: A meta-analysis. *International Journal of Educational Telecommunications* 7: 73–88.

Chapman, B. 2010. *How Long Does it Take to Create Learning?* Chapman Alliance. http://www.chapmanalliance.com/howlong/. Accessed January 12, 2015.

Chuang, C.-K., M. Chang, C.-Y. Wang, W.-C. Chung and G.-D. Chen. 2008. Application of e-learning to pilot training at TransAsia Airways in Taiwan. *International Journal on E-Learning* 7(1): 23–39.

Civil Aviation Safety Authority. 2005. *MOS Part 65 – Standards Applicable to Air Traffic Services Licensing, Version 1.2*. Canberra, ACT: Commonwealth of Australia.

Clark, R.C. and R.E. Mayer. 2011. *E-learning and the Science of Instruction: Proven Guidelines for Consumers and Designers of Multimedia.* 3rd edn. San Francisco: Pfeiffer.

Clark, R.E., D.F. Feldon, J.J.G. van Merriënboer, K. Yates and S. Early. 2008. Cognitive task analysis. In *Handbook of Research on Educational Communications and Technology*, eds J.M. Spector, M.D. Merrill, J. van Merriënboer and M.P. Driscoll, 3rd edn, 577–93. New York: Routledge.

Crandall, B., G. Klein and R.R. Hoffman. 2006. *Working Minds: A Practitioner's Guide to Cognitive Task Analysis.* Cambridge, MA: MIT Press.

Eddy, E.R. and S.I. Tannenbaum. 2003. Transfer in an e-learning context. In *Improving Learning Transfer in Organizations*, eds E.F. Holton and T.T. Baldwin, 161–94. San Francisco: Jossey-Bass.

Fabiani, M., J. Buckely, G. Gratton, M. Coles, E. Donchin and R. Logie. 1989. The training of complex task performance. *Acta Psychologica* 71: 259–99.

Fine, S.A. and S.F. Cronshaw. 1999. *Functional Job Analysis: A Foundation for Human Resources Management.* Mahwah, NJ: Erlbaum.

Fine, S.A. and W.W. Wiley. 1971. *An Introduction to Functional Job Analysis: A Scaling of Selected Tasks from the Social Welfare Field.* Kalamazoo, MI: W.E. Upjohn Institute for Employment Research.

Flanagan, J.C. 1954. The critical incident technique. *Psychological Bulletin* 51(4): 327–58.

Frederiksen, J.R. and B.Y. White 1989. An approach to training based upon principled task decomposition. *Acta Psychologica* 71: 89–146.

Gagné, R. 1970. *The Conditions of Learning.* New York: Holt, Rinehart & Winston.

Glaser, R. 1963. Instructional technology and the measurement of learning outcomes: Some questions. *American Psychologist* 18(5): 519–21.

Goevarts, M.J.B., C.P.M. Van der Vleuten, L.W.T. Schuwirth and A.M.M. Muijtjens. 2007. Broadening perspectives on clinical performance assessment: Rethinking the nature of in-training assessment. *Advances in Health Sciences Education* 12: 239–60.

Gopher, D., M. Weil and D. Siegel. 1989. Practice under changing priorities: An approach to the training of complex skills. *Acta Psychologica* 71: 147–77.

Harrow, A.J. 1972. *A Taxonomy for the Psychomotor Domain: A Guide for Developing Behavioural Objectives.* New York: McKay.

Hodge, S. 2014. *Interpreting Competencies in Australian Vocational Education and Training: Practices and Issues.* Adelaide, SA: NCVER.

Holt, R.W., J.T. Hansberger and D.A. Boehm-Davis. 2002. Improving rater calibration in aviation: A case study. *International Journal of Aviation Psychology* 12(3): 305–30.

International Civil Aviation Organization. 2006. *Doc 9868 Procedures for Air Navigation Services: Training.* Amended 2011. Montreal, QC: ICAO.

International Civil Aviation Organization. 2014. *Cabin Crew Safety Training Manual (Doc. 10002).* Montreal, QC: ICAO.

International Pilot Training Consortium. 2013. *Pilot Core Competencies: A Way Forward.* Montreal, QC: ICAO.

Jaeger, R.M. and C.K. Tittle. 1980. Prologue. In *Minimum Competency Achievement Testing: Motives, Models, Measures, and Consequences*, eds R.M. Jaeger and C.K. Tittle, v–xii. Berkeley, CA: McCutchan.

Jonassen, D.H. 1985. Mathemagenic vs. generative control of text processing. In *The Technology of Text*, ed. D.H. Jonassen, vol. 2, 9–45. Englewood Cliffs, NJ: Educational Technology Publications.

Joyner, C.W. 1995. The DACUM technique and competency-based education. In *Challenge and Opportunity: Canada's Community Colleges at the Crossroads*, ed. J. Dennison, 243–55. Vancouver, BC: UBC Press.

Kearns, S. 2010. *e-Learning in Aviation.* Aldershot, UK: Ashgate.

Kearns, S. 2013a. Recognizing the good, bad, and ugly: The four 'generations' of e-learning. *CAT: The Journal for Civil Aviation Training* 2013(1): 14–15.

Kearns, S. 2013b. Snap-courses: An instructional design strategy for aviation mobile learning. *Collegiate Aviation Review* 31(1): 69–78.

Kearns, S. 2014. Teaching with technology – Top 4 trends. *CAT: The Journal for Civil Aviation Training* 2014(4): 24–8.

Kirwan, B. and L.K. Ainsworth. 1992. *A Guide to Task Analysis.* London: Taylor & Francis.

Krathwohl, D.R., B.J. Bloom and B.B. Masia. 1964. *Taxonomy of Educational Objectives.* Handbook II, *Affective Domain.* New York: McKay.

Leung, W.-C. 2002. Competency-based medical training: Review. *British Medical Journal* 325(7366): 693–6.

Linn, R.L. and N.E. Gronlund. 1995. *Measuring and Assessment in Teaching.* 7th edn. Englewood Cliffs, NJ: Prentice-Hall.

Loevinger, J. 1957. Objective tests as instruments of psychological theory. *Psychological Reports* 3(Suppl. 9): 635–94.

Mager, R. 1962. *Preparing Instructional Objectives.* Belmont, CA: Lake.

Mansfield, B. 1989. Functional analysis – A personal approach. *Competence and Assessment* (Special Issue 1): 5–10.

Mansfield, B. and D. Mathews. 1985. *The Components of Job Competence.* Blagdon, UK: Further Education Staff College.

Mansfield, B. and L. Mitchell. 1996. *Towards a Competent Workforce.* Aldershot, UK: Gower.

Mavin, T.J. and P.S. Murray. 2010. The development of airline pilot skills through simulated practice. In *Learning through Practice*, ed. S. Billett, 268–86. Dordrecht: Springer Netherlands.

Mavin, T.J., W.-M. Roth and S. Dekker. 2013. Understanding variance in pilot performance ratings: Two studies of flight examiners, captains and first officers assessing the performance of peers. *Aviation Psychology and Applied Human Factors* 3, 53–62. doi:10.1027/2192-0923/a000041.

McClelland, D.C. 1991. Identifying competencies with behavioral-event interviews. *Psychological Science* 9(5): 331–9.

McMullan, M., R. Endacott, M.A. Gray, M. Jasper, C.M.L. Miller, J. Scholes and C. Webb. 2003. Portfolios and assessment of competence: A review of the literature. *Journal of Advanced Nursing* 41(3): 283–94.

Morrison, G.R., S.M. Ross, J.E. Kemp and H. Kalman. 2010. *Designing Effective Instruction.* 6th edn. Hoboken, NJ: Wiley.

Mosier, C.I. 1947. A critical examination of the concepts of face validity. *Educational and Psychological Measurement* 7: 191–205.

Norman, D. 2002. *The Design of Everyday Things.* New York: Doubleday.

Norton, R.E. (2004). The DACUM curriculum development process. 14th IVETA International TVET Conference (pp. 1–9). Vienna, Austria: Vocational Education and Training - International Development Catalyst

Oprins, E., E. Burggraaf and H. van Weerdenburg. 2006. Design of a competence-based assessment system for air traffic control training. *International Journal of Aviation Psychology* 16(3): 297–320.

Owen, E. and J. Sweller. 1985. What do students learn while solving mathematics problems? *Journal of Educational Psychology* 77: 272–84.

Paas, F.G. and J.J. van Merrienboer. 1994. Instructional control of cognitive load in the training of complex cognitive tasks. *Educational Psychology Review* 6(4): 351–71.

Peng, C.-Y.J. and D.J. Mueller. 2003a. Construct validity. In *The Sage Encyclopedia of Social Science Research Methods*, eds M. Lewis-Beck, A.E. Bryman and T.F. Liao, 183–4. Thousand Oaks, CA: Sage.

Peng, C.-Y.J. and D.J. Mueller. 2003b. Criterion related validity. In *The Sage Encyclopedia of Social Science Research Methods*, eds M. Lewis-Beck, A.E. Bryman and T.F. Liao, 214–15. Thousand Oaks, CA: Sage.

Piskurich, G.M. 2006. *Learning ID Fast and Right.* 2nd edn. San Francisco: Pfeiffer.

Raisinghani, M.S., M. Chowdhury, C. Colquitt, P.M. Reyes, N. Bonakdar, N.J. Ray, et al. 2005. Distance education in the business aviation industry: Issues and opportunities. *Journal of Distance Education Technologies* 3(1): 20–43.

Rosenberg, M.J. 2001. *E-learning: Strategies for Delivering Knowledge in the Digital Age.* New York: McGraw-Hill.

Rutherford, P.D. 1995. *Competency Based Assessment: A Guide to Implementation.* Melbourne, VIC: Pitman.

Sitzmann, T., K. Kraiger, D. Stewart and R. Wisher. 2006. The comparative effectiveness of web-based and classroom instruction: A meta-analysis. *Personnel Psychology* 59: 623–64.

Smith, M.V., M.C. Niemczyk and W.K. McCurry. 2008. Improving scoring consistency of flight performance through inter-rater reliability analyses. *Collegiate Aviation Review* 26: 85–93.

Smith, P. and T. Ragan. 2005. *Instructional Design.* 3rd edn. Hoboken, NJ: Wiley Jossey-Bass Education.

Stufflebeam, D.L. and A.J. Shinkfield. 2007. *Evaluation Theory, Models, and Applications.* Hoboken, NJ: Wiley.

Sweller, J. 1988. Cognitive load during problem solving: Effects on learning. *Cognitive Science* 12(2): 257–85.

Sweller, J. and G.A. Cooper. 1985. The use of worked examples as a substitute for problem solving in learning algebra. *Cognition and Instruction* 2: 59–89.

Taylor, F.W. 1906. *The Principles of Scientific Management.* New York: Harper.

Transport & Logistics Industry Skills Council. 2013. *AVI50308 Diploma of Aviation (Air Traffic Control).* Canberra, ACT: Commonwealth of Australia.

Trzmiel, B., C. Hee Kim, R. Brennan Kemmis and M. Becker. 2014. Transferable skills in technical and vocational education and training (TVET) and vocational teacher education (VTE): Policies and implementation. Editorial. *TVET@Asia* 3: 1–2.

Van Gog, T., F. Paas and J.J. van Merriënboer. 2004. Process-oriented worked examples: Improving transfer performance through enhanced understanding. *Instructional Science* 32: 83–98.

van Merriënboer, J.J.G. 1997. *Training Complex Cognitive Skills: A Four Component Instructional Design Model for Technical Training.* Englewood Cliffs, NJ: Educational Technology Publications.

van Merriënboer, J.J.G., P.A. Kirschner and L. Kester. 2003. Taking the load off a learner's mind: Instructional design for complex learning. *Educational Psychologist* 38(1): 5–13.

van Merriënboer, J.J.G., R.E. Clark and M.B.M. de Croock. 2002. Blueprints for complex learning: The 4C/ID-model. *Educational Technology, Research and Development* 50(2): 39–64.

Wickens, C.D., S. Hutchins, T. Carolan and J. Cumming. 2013. Effectiveness of part-task training and increasing-difficulty training strategies: A meta-analysis approach. *Human Factors* 55(2): 461–70.

Woehr, D.J. and A.I. Huffcutt. 1994. Rater training for performance appraisal: A quantitative review. *Journal of Occupational and Organizational Psychology* 67(3): 189–205.

Zhao, Y., J. Lei, B. Yan, C. Lai and H.S. Tan. 2005. What makes the difference? A practical analysis of research on the effectiveness of distance education. *Teachers College Record* 107(8): 1836–84.

PART IV
The Future of Competency-Based Education in Aviation

Introduction to Part IV
The Future of Competency-Based Education in Aviation

Part IV of this book provides concluding chapters on the future direction of competency-based education (CBE). Before moving forward, it is important to reflect upon the definitions of CBE concepts that have been discussed throughout this book:

Competence: the ability to fully participate in a complex social practice, such as an aviation profession. Full participation requires skills, knowledge and attitudes relevant to that practice.

Competencies: negotiated and agreed written statements (texts) that attempt to represent the ability to fully participate in a social practice. Competencies embody assumptions about:
- the nature of competence
- the number of discrete written statements (texts) required to represent competence
- the most appropriate language for representing competence.

Competency-based education (CBE): instructional design, training and assessment that systematically references written competencies.

This definition takes into account competence, competencies and competency-based education. For a graphic that summarizes these concepts, please see Figure P1.1 in the Part I introduction of this book.

For **competence**, please consider that:

- Competence refers to the ability to effectively function in a professional role.
- Competence is **what a professional actually does** while on the job.
- Competence is not the same as expertise. It can be expected that a learner who has achieved competence will require many additional years of experience on the job to develop expertise. For example, consider the difference between a "competent" pilot who has just earned his or her airline transport pilot license, and an "expert" airline captain.

For **competencies**, please consider that:

- Competencies refer to written statements (texts) that have been created to describe professional abilities.
- Competencies *try* to describe, in written language, what a professional actually does while on the job.
- It is critical to recognize the difference between "competence" and "competencies." As soon as a person puts pen to paper, competence (all of the abilities included in expert performance) will be simplified. It is very difficult to translate expert performance into words, and therefore **it is possible for a gap to exist between true "competence" and the written "competencies."** Competencies can be reductionist— meaning they are an artificially simplistic representation of competence.
- A large amount of negotiation often takes place before agreement is reached on the language used in competencies. Remember that consensus-based approaches don't always reflect what is ideal; they reflect what could be agreed upon by a group of experts who sometimes have differing views.
- When competencies are agreed upon and distributed, different people interpret them in different ways. Understanding written competencies requires a skill in itself, called "hermeneutics," which refers to a person's proficiency in interpreting text. If aviation does not teach hermeneutics as a component of CBE, even the most carefully crafted competencies could have wide variability in their interpretation around the globe.

Competency-based education (CBE) is instructional design, training and assessment that systematically references competency texts:

- "Competency-based education" is instruction that is created to focus learning activities on the development of professional "competence." This is accomplished by linking instructional objectives to "competencies" (textual written statements).
- Therefore, compared with traditional training, competency-based education is more focused on **application** of learning to realistic scenarios. However, the majority of traditional instructional practices still play a vital role within competency-based education; they are typically expanded upon to ensure that learners can *apply* what they have learned. An issue the authors noticed in interviews is that some advocates of CBE vilified and criticized traditional practices. In reality, CBE is more similar to than different from traditional instructional approaches.

Part IV: Chapter Descriptions

Chapter 10: Instructional Technologies in Aviation Training: Today and Beyond: instructional technologies are evolving rapidly and are expected to have a large impact on future training practices. This chapter explores the impact of simulation and discusses how future professionals may approach learning, big data, wearable technology, augmented reality and adaptive e-learning.

Chapter 11: The Future of Competency-Based Education in Aviation: this chapter explores some of the strengths and weaknesses of competency-based education in aviation and indicates ways the industry can move beyond these weaknesses.

Chapter 10

Instructional Technologies in Aviation Training: Today and Beyond

Questions Considered in this Chapter:

- What is the role of instructional technology for the aviation professionals of today?
- How will future aviation professionals learn?
- How might we expect technology to evolve in coming years?
- How will future technologies impact the personal lives, workplace and learning experiences of aviation professionals?

Introduction

Aviation training has long used an instructional approach that blends strategies: a combination of classroom-based theoretical instruction and real-world practical training. As an industry that requires continual recurrent training throughout a professional's career, aviation has a history of being an early adopter of new teaching technologies that seek to improve efficiency or safety, or to lower costs.

Simulators were developed within aviation, with the first introduced around 1910, and in the mid-1980s our industry was one of the first professional fields to embrace computer-based training (Kearns 2010). Technology is now evolving at a rapid pace, and it is reasonable to expect that new teaching technologies will quickly make their way into aviation professionals' curricula and reshape training in the future.

The goal of this chapter is to review how technology has shaped aviation training (through a short history of flight simulators and their use within today's aviation training system), explore some examples of rapidly evolving technologies and discuss their potential application to future aviation training.

A word of caution: although the ideas presented in this chapter propose technology-based modifications to training programs, we must remain mindful that new technology does not necessarily result in more effective training. Before technology is implemented, it must be evaluated to ensure that it leads to training improvements, rather than being impressive but ineffective.

Technology in Today's Aviation Training: The Role of Flight Simulation

When considering technology in aviation training, it is tempting to jump immediately to present-day digital marvels. However, it is crucial to first consider the accomplishments of the past, which show that "the driving force of technology evolution is not mechanical, electrical, optical, or chemical. It's human: each new generation of simulationists standing on the shoulders—and the breakthroughs—of every previous generation" (Adams 2010, para. 1).

Partly because of the demand for pilots generated by the World Wars, it wasn't long before the aviation industry began struggling with how to safely teach courageous trainees as safely as possible. This was complicated by the fact that the earliest aircraft had "no static stability, so you have to constantly stay on top of it. It can suddenly go off in any direction as it responds to even the slightest breeze or thermal" (Terry Queijo, quoted in Ulrich 2010, para. 3). Unfortunately, fatal and hull-loss accidents were a frequent occurrence, which led to some interesting ideas associated with how to teach flight at low altitudes or on the ground so there would be less risk to learners and aircraft. Early experiments in flying instruction included "grass cutting" at half throttle (flying at very low altitudes to practice straight flight), altered pitch propellers that allowed aircraft to jump but otherwise stay land-bound, and reduced wingspans to allow practice without becoming airborne (Ulrich 2010).

The earliest aviation simulators were introduced around 1910:

- the Wright Brothers' "kiwi bird," which was a defunct Wright Type B Flyer mounted on a trestle
- the Antoinette Trainer, in which learners would sit in a barrel while instructors pulled or pushed the "wings" (to simulate disturbances), requiring the learner to make control inputs through a series of pulleys (Moroney and Lilienthal 2009).

However, it wasn't until 1929, with Edward A. Link's trainer with three degrees of freedom (yaw, pitch and roll), that modern flight simulators were born. The Link Trainer used compressed air to simulate flight movement and allow learners to "feel" the impact of stick inputs on flight characteristics (Moroney and Lilienthal 2009).

From modest beginnings, modern flight simulators have evolved to incorporate cockpits with controls and displays nearly identical to those of the aircraft themselves, realistic performance, high-fidelity visual imagery representing the outside world, motion and vibration, sounds (including human communication and effects), malfunction capability and full instructor controls (G. Woolley, personal communication).

The International Civil Aviation Organization (ICAO) defines a *flight simulator* as follows:

> A full-size replica of a specific type or make, model and series of aeroplane flight deck, including the assemblage of equipment and computer programmes necessary to represent the aeroplane in ground and flight operations, a visual system providing an out-of-the-flight-deck view and a force cueing motion system. It is in compliance with the minimum standards for flight simulator qualification. (ICAO 2003, 1–2)

Modern simulators are classified based on their *fidelity*, that is, how similar they are to real-world operations and the types of failures and effects they can accurately produce. As the goal of this chapter is to provide a general description of the role of simulators in aviation training, it will not include details on how simulators are categorized or regulated, or specific training credits associated with types of devices (for more information see Adams 2009).

Figure 10.1 presents a progression of simulators from basic desktop e-learning through to full mission simulators (FMSs). Each level of the progression represents a higher-fidelity simulation device than the previous, but each has its place within a training curriculum. As higher-fidelity devices are associated with higher purchase costs, it is strategic to use the lowest fidelity device that will accomplish the training objective. For example, if the goal is to practice procedures, it is more efficient to use a less expensive flight navigation and procedures trainer (FNPT) than a more expensive full flight simulator (FFS). Furthermore, there is a group of researchers who argue, quite reasonably, that the relationship between learning and fidelity is not linear (e.g., Roscoe 1991). An increasing amount of discussion—not only within aviation but also in other professions such as medicine—also suggests that for already trained professionals, simulation fidelity and development of expertise is not as clear as one would imagine:

> Two potential uses of [high-fidelity simulators] may lie in the teaching of an approach to rare and difficult problems, and in the teaching of the management of complex problems involving multiple health professional roles. Paradoxically, neither may require any sophistication of simulation. In the context of a rare problem, experts may regard the issue as a cognitive problem and simply "think" their way to a solution. Their experience in the real setting is likely to be sufficiently extensive to make practice of the motor skills involved superfluous. In the context of a complex problem that requires input from multiple health professionals, the focus of activity is between professionals; the features of the simulator may be irrelevant. (Norman, Dore and Grierson 2012, 645)

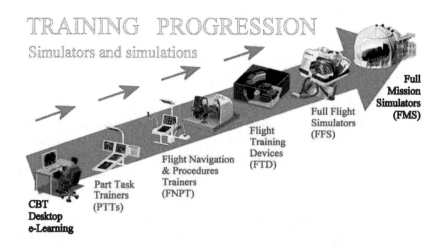

Figure 10.1   Training Progression of Simulators and Simulations
Source: Ian Strachan.

The following are examples of activities that could take place in the various levels of the progression:

- **Computer-based desktop e-learning.** These platforms are often used for engineers learning initial systems knowledge such as electrical systems or hydraulics. They can be linear in delivery or in some cases allow students to move into areas of interest.
- **Part-task trainer (PTT).** In almost all modern aircraft, pilots are required to follow a particular flow of procedures for all stages of flight; many instructors will refer to these as "your flows" or "your scans." The PTT allows pilots to practice their flows, which in almost all cases will require a sequence of actions that interact with the second pilot. This flow has been likened to an orchestra, with different parts coming together as a single melody. In fact it has been referred to as "kinetic melody" (see Roth, Mavin and Munro 2014). The PTT is an economical simulator that allows this practice to occur. In fact, in years gone by, the PTT could be as simple as a poster of the aircraft's flight instrument panel and switches, with the pilots touching each instrument/switch/dial in the required order.
- **Flight navigation and procedures trainer (FNPT).** A pilot/engineer may be required to initially learn the use of a flight management computer, and the only skill to be learned might be how to load a flight plan or make cruise flight changes. Here the pilot/engineer will only have a replica of the flight management computer and a monitor to show flight parameters.

- **Flight training device (FTD).** The flight training device is a simulator that replicates flight. Depending on the need of the training organization, the FTD may be a *generic* mockup of an aircraft, displaying primary flight instruments and system switches or dials. Or it can be an *identical* representation of the aircraft the pilot will fly. FTDs generally do not have motion capability, although sometimes "seat movement" can be fitted to simulate turbulence.
- **Full flight simulator (FFS).** Full flight simulators are now commonly used to train experienced flight crew in a zero flight time (ZFT) approach, meaning that all training is in a simulator and the first time the crew member operates a new type of aircraft there will be revenue-generating passengers in the back. The FFS replicates motion (though only very new, highly advanced simulations in production can replicate G-force), with a flight instructor acting as other aircraft, air traffic control, engineering and cabin crew.
- **Full mission simulator (FMS).** Full mission simulators, which are used especially in military contexts, attempt to replicate the realities of actual missions. They have either a high-fidelity FTD or FFS that replicates the aircraft type the pilots fly and can also be linked with other FTDs or FFSs. For example, two fighter pilots may be in two different FTDs, but will fly as a pair during a combat mission. In this case, the pair maybe required to fight enemy aircraft being "flown" by an instructor in another room.

According to a 2014 flight simulator census, there were 1,429 FFSs, 78% of which were at the highest level of fidelity. The majority of these (66%) were located in North America and Europe. Asia and the Middle East are high demand markets for new simulators, with Asia expected to need 970 and the Middle East requiring approximately 200 new devices in the coming years (Strachan 2014).

However, aviation simulation is far from limited to flight crew applications. As described by the Royal Aeronautical Society Flight Simulation Group (2009), sophisticated and high-fidelity devices are used within most aviation professional groups:

- Automated air traffic control simulators represent highly complex visual and communication systems that generate realistic environments, many with speech recognition capability.
- In the military flight training environment, simulators allow for full mission rehearsal that would otherwise be prevented by strategic and tactical secrecy (such as practicing an operational sortie over unfamiliar territory).
- Aviation maintenance personnel are able to practice fault-finding, removal and installation of equipment in a simulated environment without the use of actual parts.
- Aircraft cabin simulators allow cabin crew to practice emergency procedures.

Digital simulations are increasingly popular for developing affective and complex cognitive skills in all areas of professional training (Gegenfurtner, Quesada-Pallares and Knogler 2014; Tennyson and Jorczak 2008). As simulator technology has evolved over the years, attention has focused on issues of instructional methodology (such as scenario-based or team training), visual fidelity and whole-body motion (de Winter, Dodou and Mulder 2012; Salas and Burke 2002). Overall, simulators are generally regarded as an effective and environmentally friendly alternative to real-world training that results in saved dollars, materials and even lives (Moroney and Lilienthal 2009; Sitzmann 2011).

It is expected that simulator-based training will continue to evolve. Possible future developments might include (1) downloading of pre-existing scenarios from the cloud, (2) adaptive and data-driven training based on algorithms that analyze learner characteristics and organizational needs, and/or (3) delivery through less expensive head-mounted augmented reality (AR) systems.

For a review of competency-based classroom teaching methods, including e-learning, please see Chapter 8.

Technology's Role in Future Aviation Training

How Will Future Professionals Learn?

Before we discuss future teaching technologies, it is important to consider how future professionals in our industry may approach learning. As the next generation of aviation professionals enter our industry, our training systems may have to adapt to align with their learning preferences and styles. Prensky (2001) suggests that today's students are no longer the people our training system was designed to teach. He explains that younger generations have a new approach to learning as a result of growing up immersed in technology. This can be understood through the term *digital native*, referring to the younger generation who have grown up with technology as a part of their daily lives, as opposed to *digital immigrant*, meaning the older generation who were required to learn to use technology as an adult. Digital natives have a fundamentally different approach to thinking as well as processing information. Compounding this challenge, older digital immigrant instructors may speak an outdated language and struggle to relate information to digital natives. Overall, "digital immigrant teachers assume that learners are the same as they have always been, and that the same methods that worked for the teachers when they were students will work for their students now. **But that assumption is no longer valid**. Today's learners are *different*" (Prensky 2001, 3).

Smith (2012) has identified eight characteristics associated with how digital natives learn:

1. They have a new way of knowing and being.
2. They are driving a digital revolution that will transform society.

3. They are inherently tech-savvy.
4. They naturally collaborate and multitask, and are team-oriented.
5. They can be considered "native speakers" of the language of technology.
6. They embrace simulation, gaming and interaction.
7. They expect immediate gratification.
8. They demand knowledge in the Information Age.

The digital native concept is an important consideration in designing training. However, it is not panic-worthy; although young learners prefer technology, and use more of it, they also seem to conform to the approach used by their instructor (Margaryan, Littlejohn and Vojt 2011).

Beyond how digital natives learn, another (often more frustrating) issue is learners becoming distracted by their personal devices. Current educators only have to take a quick glance around a classroom to realize that the younger generation are continuously connected to the online world. This connectivity poses a significant distraction in a learning environment. Students distracted during class receive 11% lower exam grades than their peers. Interestingly, these distractions are not limited to the individual learner using the device. Fellow learners in a classroom who were not using technology yet were seated near a web-using student scored 17% lower on their exams—demonstrating that distraction from others poses a more negative impact on learning than using one's own device (Sana, Weston and Cepeda 2013). It is difficult for even the best instructors to compete with technology, as learners are only a click away from a world of information. Perhaps instructional approaches that integrate teaching technologies would be more engaging for digital native learners. One application of technology to the classroom is based on the collection and use of a large number of individual and organizational data points, called *big data*.

Big Data

Cisco (2015) estimated that by 2015 there would be 5.2 billion mobile users around the globe, up from 4.3 billion in 2014. The increasing popularity of these devices generates an enormous and ever-increasing amount of data. Mobile data traffic in 2014 was estimated at 4 zettabytes, almost 30 times the size of the entire global Internet in 2000 (Cisco 2015; Pentecost 2015). To put that into perspective, if your large cup of coffee equaled one **gigabyte**, a single **zettabyte** would have a volume equal to the Great Wall of China (Cisco 2014).

To understand where all the data is coming from, it is helpful to consider the features of a mobile device. Mobile devices incorporate a range of data-generating components:

* global positioning system (GPS)
* camera
* keypad
* web-connectivity

- phone
- step and speed trackers.

As devices are now a vital part of our personal and professional lives, each of us is generating a stream of data that can be thought of as digital breadcrumbs. Just as Hansel and Gretel left a trail of breadcrumbs as they walked down the forest path in the Brothers Grimm fairy tale, as we move about during our day we generate data points associated with where we were, who we spoke to, what information we referenced and anything we documented.

Beyond the data generated by individuals, a massive amount of data is created and recorded by machines. For example, utilities track household and commercial energy usage to predict outages and incentivize conservation, Google uses the number of illness-related web searches to detect disease outbreaks and general retailers track clicks on their webpages to better understand consumer behavior and improve sales (MongoDB 2014).

On a global scale, this has created a new industry that seeks to generate, analyze and sell insights from the massive pool of big data. For example, algorithms that analyze big data can allow companies to market more effectively by targeting specific consumers who are most likely to make a purchase or to predict how politicians will vote based on their history at all previous levels of government (*The Economist* 2014).

How Will Big Data Impact Aviation Training?

Some interesting applications of big data are in use within the aviation industry. One usage is associated with proactive safety management systems within an airline, where cluster analysis is used on routine operational data from flight data recorders (FDRs) to identify anomalies and associated safety hazards. This allows the airline to identify airports where early signs of performance deviations are occurring, identify what safety hazards are present, predict and implement safety remediation, and train staff accordingly (Li and Hansman 2013). This is a tremendous step forward, as traditional safety management systems rely on personnel reports or the number of incidents that have occurred. This shift presents a truly predictive approach to safety that identifies risks "beyond the capacities of current methods" (3).

Etihad Airways is already incorporating big data to compare data from its operations and training sectors (e.g., use of rudder, crosswind landings, rejected take off actions, rate-of-rotation or flare, etc.). This comparison of the accumulated data from simulated and line operations scenarios:

> will lead to a better understanding of exactly what training is required, which falls neatly into the integration of Evidence Based Training, which in turn is coupled with Competency standards. Not only will the average performance be established, but individuals will be able to see and understand where their own performance is relative to the fleet and the acceptable norm. (Long 2014, 8).

Beyond operations, big data would also contribute to recruitment and selection practices by allowing organizations to use predictive analytics to compare job applicant performance against competency demonstrated by top performing employees. Once a professional has joined a company, machine-driven learning algorithms could continually analyze that person's performance data, alongside operational statistics, to detect areas where a shortcoming or weakness exists and to assign supplementary training as required.

These data-driven predictive practices would impact the type of professionals hired into a company, in addition to customizing the training completed throughout an individual's career. These insights can lead to customizations in the training received by each individual, based on organizational needs and learner characteristics.

Challenges of Big Data

Although there are advantages to big data, both to industry and consumers, there are also risks. The continual collection of data can lead to privacy breaches, abuse of personal data or even cybercrime. Are individuals willing to share personal data in order to receive innovative services? How can this be effectively regulated and overseen? Young people are quick to take advantage of technology and embrace having the world at their fingertips, yet they may not realize that they put themselves at the world's fingertips as well (Crowther 2012). Evidence suggests that the next generation of professionals care about their privacy, but they lack knowledge of privacy issues (Hoofnagle et al. 2010). Perhaps this is because they are immersed in an online reality that is designed to encourage them to reveal personal data.

Big data is far from perfect. It is only as accurate as its source information. For example, compared with slower, more traditional data collection processes, Google Flu Trends significantly overestimated the flu outbreak (Butler 2013). And in the process, Google broke a wall of trust by accessing users' private searches in a manner that goes beyond what are historically accepted uses. In addition, Google has given researchers access to our searches and "a motive to focus in particular on some of the most sensitive information about us, our medical symptoms" (Ohm 2012, 342).

Although obvious privacy concerns exist, those who advocate big data suggest two main benefits: (1) predictive analytics as a way to forecast future behaviors, actions or even accidents before they occur; and (2) constant learning through machine-driven learning algorithms to continually explore big data sets and identify associations (Pentecost 2015).

As mentioned previously, the quality of findings from big data is entirely reliant on the quality of information received. A potential source of insight into individual differences is data from wearable technology.

Wearable Technology

Wearable technology refers to portable devices that attach to the human body, collecting data and delivering information. Although a very new consideration in aviation training, these devices are already prolific, as there were nearly 109 million wearable devices around the globe in 2014 (Cisco 2015). The idea of symbiosis between humans and technology is not new: it has been around for at least 50 years (Licklider 1960). Yet it is only in recent years, with technology becoming smarter and more portable, that it has become a true possibility.

Mobile telephones revolutionized human social and technological interactions, yet wearable devices (which possess many of the same features as mobile phones) take this further, through the integration of *scanning features* and *sensors*. The data collected by scanning features and sensors have a wide range of applications. Examples include the following:

- stress-sensing gloves that predict anxiety levels (Picard and Scheirer 2001) and acceleration-sensing gloves (ASG) that determine movement (Perng et al. 1999)
- washable "smart shirts" that sense health data (respiration, heart rate, electrocardiogram, body temperature and pulse oximetry) for civilian health monitoring, or to detect trauma of soldiers on the battlefield (Park and Jayaraman 2003)
- cyberclothes, such as pants, that are outfitted with accelerometers that sense when a person is walking, running, climbing or descending stairs, or riding a bicycle (Duval and Hashizume 2005; Van Laerhoven and Cakmakci 2000)
- fitness wristbands that measure heart rate, oxygen consumption, sleep and movement (such as steps taken in a given day; Wright and Keith 2014)
- jewelry, such as hearing aids that are designed to look like earrings or bracelets that can transmit an emergency distress signal (Wright and Keith 2014)
- hazardous-situation smart suits that intelligently sense a wearer's surroundings and suggest appropriate actions. (Rogers, Murphy and Thompson 1997)

Wearable technology is entering a rapid growth phase, with the market expected to reach $70 billion by 2025 (Harrop et al. 2015).

How Will Wearable Technology Impact Aviation Training?

The effectiveness of instruction will always be linked to a number of variables associated with a learner's physiological state. For example, did they get sufficient sleep, are they impaired by substances or medications, are they stressed or overwhelmed by information in a classroom, are they looking at the right instrument during a simulator exercise, did they exert sufficient force on the control or tool, or are they physically overworked?

While skilled instructors and mentors often understand and recognize these issues in learners, wearable technology has the potential to track this type of data automatically and to objectively present instructional recommendations about how to tailor instruction.

The possibilities for use of wearable technology in aviation are as broad as one's imagination. For example, early devices that track eye movements and measure electrodermal activity (EDA) have the ability to sense different types of stress. That is, they can distinguish work-related social or time-related stress from stress associated with cognitive load, which occurs when a learner is becoming overwhelmed by information (Setz et al. 2010). This suggests that devices in the future may help instructors determine when a learner has achieved competence: the learner who has achieved such competence may no longer exhibit cognitive load stress in challenging scenario-based training.

Challenges of Wearable Technology

The three main challenges associated with the integration of wearable technology are privacy, adoption and cost.

Privacy can be understood through two basic perspectives: Warren and Brandeis's (1890) definition of privacy as *the right to be left alone,* or Westin's (1967) data-protection approach that focuses on *regulating when, how and the purpose for which data is collected, used and disclosed* (Iachello and Hong 2007). How can the opportunities offered by technology be adopted without sacrificing privacy? An evaluation can be conducted that considers three key factors:

1. *adequacy:* ensuring there is a balance between enough and not-too-much information collected
2. *appropriateness:* the correct technology is used from those available
3. *legitimacy:* determining if the goal of data collection is useful and beneficial. (Iachello and Hong 2007)

Wearable technology is capable of transmitting a range of private data, such as location, personal information and communications, and health data. Wearers are likely to have concerns surrounding the accidental publication of private information, whether the data collected could be unlawfully or unfairly used by others, and infringements of the wearer's privacy and the privacy of people nearby (Kirkham and Greenhalgh 2015). Consider how aviation professionals might perform differently if they knew that "big brother" (their company, insurance agency, regulator or the general public) were continually watching their performance? How would unions be likely to respond?

A related challenge surrounds whether or not professionals would be willing to adopt these devices. Although previous work associated with mobile phones demonstrated that individual views of technology shift after a person acquires a device—that is, nonusers who disliked mobile devices changed their mind and enjoyed mobile smartphones once they acquired a device (Palen, Salzman and Youngs 2001)—there is a high potential for individuals to object to continual monitoring of personal data.

A final issue surrounds the high cost associated with these devices. Cost is the most common barrier (72%) to acquiring devices (Nielsen 2014). Although prices typically decrease as the technology improves, the niche-use of individual devices for specific tasks may remain a challenge, as each person would require a range of devices.

Overall, wearable technology is associated with many broad issues, including possible instructional applications as well as serious challenges. However, one type of wearable device has significant potential within aviation training: wearable head-mounted optical displays to facilitate augmented reality.

Augmented Reality

As Azuma (1997) explains, augmented reality (AR) is a variation of virtual reality (VR). VR is best understood as the approach used in modern flight simulation devices, where a learner sits in a mock flight deck and the outside world is presented on a video screen. While VR immerses a user within a synthetic environment, eliminating their ability to see the real world around them, AR "allows the user to see the real world, with virtual objects superimposed upon or composited within the real world" (356).

According to Azuma et al. (2001, 34), an AR system has "the following properties:

- Combines real and virtual objects in a real environment;
- Runs interactively, and in real time; and
- Registers (aligns) real and virtual objects with each other."

Optical head-mounted displays, such as Google Glass, are an example of wearable technology that supports AR. Google Glass technology incorporates a bone-mounted transducer to transmit audio to the wearer, microphone for dictation and communication from the wearer, video camera, gyroscope, GPS, Wi-Fi, Bluetooth (including Internet connection), and a display to visually present AR data (Vallurupalli et al. 2013).

Modern AR incorporates machine learning algorithms to "intelligently" sense the environment and support the individual in the task. For example, within a hazardous environment AR program, an *attention director* component highlights key aspects of the environment that must be attended to, a *task director* organizes the massive amount of data in the environment, and a *hypothesis manager* reviews all of the data and compares expectations to observations to make a guess about what is occurring in the situation (Rogers et al. 1997).

How Will Augmented Reality Impact Aviation Training?

Students using AR systems in a classroom could reference textbooks, media or systems and components while simultaneously listening to an instructor's presentation.

Google Glass has been successfully integrated into simulation-based training and debriefing in the medical field. It has primarily been used to record a learner performing a simulator exercise, so that the recording can be replayed and analyzed during debriefing (Vallurupalli et al. 2013; Wu, Dameff and Tully 2014).

Beyond classroom use, a variety of applications of AR systems can be applied to the real-world workplace of aviation professionals. Each of these applications would require training on the proper use and application of the devices. For example, AR could be used to:

- replace air traffic controllers' paper flight strips with a digital presentation of the information superimposed on their work space (Mackay et al. 1998)
- present maintenance personnel with a digital view of systems that may not be visible because they are obstructed by cowls or other components (Sims 1994)
- display the name and preferences of passengers to cabin crew as they walk through the cabin
- generate a digital representation of the ideal approach path for pilots as they begin their descent to land.

Challenges of Augmented Reality

Although AR technology shows great promise, it has several drawbacks. For example, most AR devices are very cumbersome (some complain that they "look goofy") and have a short battery life that limits their operational applications.

In January 2015 it was announced that Google Glass would no longer be produced. Although the initial launch was considered a successful experiment, Google claimed it was going back to the drawing board to evolve the technology (BBC 2015). While Google rethinks their technology, there are several other smart glass developers on the market (for example, Vuzix Corporation, which produces optical devices for the United States military; Evena Medical, the developer of Eyes-On Glasses; and Epson's AR glasses) and it is expected that this technology will continue to evolve (Wright and Keith 2014).

Intelligent Tutoring Systems (ITS)/Adaptive E-Learning

Beyond the use of big data to identify training needs within a company, and wearable technology to provide learner data to instructors, data can also drive the structure of a training curriculum. Traditional classroom and computer-based training targets the needs of the average learner, the 50th percentile. This means that instruction will be too easy and possibly boring for high-performers (experts) and too complicated for underperformers (novices). In addition, classroom instruction is usually delivered in an instructor-centric (as opposed to learner-centric) style. The differences between instructor- and learner-centric instruction can be understood through the information in Table 10.1.

Table 10.1   Comparison of Instructor-Centric vs. Learner-Centric Characteristics

Instructor-centric teaching	Learner-centric teaching
Knowledge is delivered from instructors to learners	Learners build knowledge by gathering, compiling and integrating new information using critical thinking, inquiry and problem-solving skills
Learners receive information passively	Learners are actively engaged and involved
Emphasis on memorizing knowledge	Emphasis on using, communicating and applying knowledge to real-life situations
Instructor is the primary deliverer of information and primary assessor	Instructor is the coach and facilitator; instructors and learners evaluate learning together
Instructing and assessing are separate	Instructing and assessing are interwoven
Assessment is used to measure learning	Assessment is used to diagnose and promote learning
Assessment based on learners knowing the "right answers"	Assessment based on learners learning from mistakes and understanding the complete situation
Usually focused on a single topic (air law, meteorology, navigation, etc.)	Usually incorporates interdisciplinary and whole-task learning
Culture among learners is individualistic	Culture among learners is supportive, collaborative and cooperative

Source: Adapted from Huba and Freed 2000.

The goal of most training is to have learners pass a test or assessment to prove that they have reached the required threshold of learning. Using student-centric methods to facilitate each individual reaching their highest possible level of knowledge and performance would not be logistically possible with a single instructor and a limited training budget.

Yet, as technology evolves, new courseware is emerging that is designed for this purpose—to *adapt* to the needs and learning style of each individual learner so that they can achieve the highest level of learning possible based on their unique intellectual and learning capacity. This approach to learning has several names, including intelligent tutoring systems (ITS), adaptive e-learning, or level 4 e-learning (see Chapter 8 for a development cost estimate for Level 4 e-learning).

E-learning can be categorized as either *static* or *adaptive*. Static learning material is created in a specific sequence designed by an instructor (instructor-centric). Learners complete the training in a linear fashion, beginning with the first unit of training and working their way through to the end. Adaptive learning (student-centric), such as an ITS, broadly refers to any educational computer program that utilizes some type of artificial intelligence to guide the structure of the curriculum. The complexity of adaptive projects varies tremendously, depending on the level of artificial intelligence and student-centric methodologies that are integrated (Freedman 2000).

Whereas the majority of research into e-learning explores multimedia characteristics (the types of pictures, animations and audio that optimize learning; see Chapter 8), exploration of personalization and intelligent tutoring is a relatively new field that combines education research, artificial intelligence and Internet technology (Cheung et al. 2003). ITSs combine these attributes to give personalized advice to students based on their background and abilities, and the needs of their professional role.

How Does Adaptive E-Learning Work?

The artificial intelligence component of adaptive learning drives its ability to customize training content to learner needs and preferences. This is accomplished by having the system *learn* about each person and generate content through the following components:

Course Manager

1. Student model
2. Automatic curriculum generation
3. Metadata
4. Expert model
5. Question bank (Cheung et al. 2003).

The *course manager* has the overarching role of analyzing and incorporating all of the other structures of the software. This will include consideration of the student model, content and structure of courses, and the expert model (described in the following paragraphs). The ultimate job of the course manager is to create a tailored course that references the external training needs of an organization, gives individualized instructions and suggestions during the learning process, maintains student records (personal information, learning status and testing results), delivers the course through the user interface and creates adaptive assessments (exams) via the question bank. In short, the course manager coordinates all of the ITS components so that training is effective and the learner's experience is smooth and precise (Cheung et al. 2003).

The *student model* stores information about each individual learner. This data is gathered with every interaction between a specific learner and the ITS platform. Typically, the data is associated with

- *Learning preferences:* What type of media is preferred? Does this learner benefit from interactive exercises? What level of difficulty should be presented?
- *Cognitive state:* What does the learner know already? (Capuano, Marsella and Salerno 2000)

The *automatic curriculum generation* process is responsible for identifying a unique curriculum for each learner, based on their identified needs and the goals of the organization. Features include:

- *Learning goals:* The system will identify "what has to be learned" by this specific individual for them to reach competency.
- *Curriculum:* A unique list of learning objects will be compiled to accomplish "how it will be taught."
- *Learning objects:* Specific learning activities that suit the individual's learning style will be created to support the "activities the learner interacts with—the lesson itself." (Capuano et al. 2000)

In order to adaptively select the appropriate learning object, all learning objects are "tagged." This indexing of learning objects is referred to as *metadata*. Metadata is information used to efficiently locate digital resources within a computer program; it describes the learning object and its relationship to other educational material. This creates a very complicated flow chart that describes the prerequisites, subconcepts and relations of every learning object (Capuano et al. 2000).

Similar to the process of identifying competencies, modeling the knowledge learners need within an ITS is a complicated process that needs to consider multiple types of expertise. This is called the *expert model* and includes a professional's subject matter knowledge; a system-created learner model to track each individual's ability, perceptions and actions; and pedagogical expertise about how to teach most effectively. All of these types of expertise evolve continuously and ITSs must adapt to each learner and different learning situations (Frasson, Mengelle and Aimeur 1997).

Lastly, a large *question bank* is required within an ITS. Question banks become very large because each question must be written several times, at multiple levels of difficulty, to target unique learning levels. For example, a single assessment may be written at novice, intermediate, expert and mastery levels of difficulty. Each of these four levels may also have different versions to target learners with different learning styles.

Challenges of Adaptive E-Learning

As discussed in Chapter 8, the development costs associated with Level 4 adaptive e-learning are significantly higher than for other types of e-learning courseware. Because of these high costs, this approach to training might not be feasible for content areas that evolve (such as regulations or company procedures that may be updated regularly). Within aviation training, ITS is best applied in content areas that have relatively static information and that are usable by the largest possible population, as large student numbers justify higher development costs.

Although adaptive e-learning has its advantages, in some ways it has disadvantages compared with a human tutor. Human tutors can intuitively sense when a learner is overwhelmed or becoming frustrated and can adapt instruction as necessary. Based on the premise that learning is enhanced when instruction takes a learner's emotions into account, ITSs are working to enhance the digital learning environments so they can adapt to the learner's emotional state through emotion-sensing technology. Technology can effectively measure facial expressions and body movements through a device's webcam and track conversational cues through the microphone. This data can be analyzed to determine if a learner is becoming frustrated, confused or bored (D'Mello et al. 2005).

Another challenge is that sometimes ITSs can be "gamed" by learners, meaning that students exploit the features of the system rather than learning the content of the educational material. This problem can be reduced by introducing an animated avatar character that expresses negative emotions to gaming students and by providing supplementary exercises on any material bypassed by gaming (Baker et al. n.d.).

With new technology, questions arise as to the effectiveness of the adaptive e-learning approach. Will there be a return on investment for this type of courseware? Early research in this area is promising. Wickens et al. (2013) compared the effectiveness of whole-task training that increased in difficulty at a *fixed* schedule with training that incorporated an *adaptive* schedule (where the difficulty increased for learners adaptively based on their performance) and they found a clear benefit for adaptive learning schedules. Adaptive learning produced a 36% benefit compared with traditional learning in which the difficulty level was static and set to the average learner's ability. However, in a course where the difficulty increased at a fixed schedule (not related to the learner's abilities) the benefit was reversed and resulted in a 23% reduction in learners' performance compared with traditional learning. Therefore, to be more effective than static courses, adaptive e-learning must incorporate artificial intelligence that adapts according to the student model (learner-centric). If the adaptive elements adjust at a fixed schedule (instructor-centric), the course ends up being less effective than a static e-learning course.

How Would This Work in Aviation?

Interestingly, one of the first practical ITSs was developed within aviation. It was a computer-based Air Force electronics practice tool called "Sherlock" (Lesgold et al. 1988). A training challenge for the U.S. Air Force in the mid-1980s was that F-15 avionics technicians filled a position for a brief period (typically four years or less) and mostly carried out routine tasks that were well supported by technology. These routine, semiautomatic jobs did not allow technicians the opportunity to develop complex problem-solving skills. This left the Air Force with a problem: training assumed that test station repair system troubleshooting would be learned on the job, but the job did not offer enough practice opportunities. Sherlock was created as an environment where avionics technicians could practice troubleshooting skills and it was extremely successful. Novice technicians who practiced on Sherlock for 20–25 hours developed troubleshooting skills comparable with those of their colleagues with four years of on-the-job experience (Lesgold et al. 1988).

Sherlock incorporated hints to support the trainee's efforts in solving a fault isolation problem. Some hints were dynamic, as in response to an incorrect action, while most were canned; for example, in response to the trainee hitting a panic button, Sherlock would present a top-down overview of portions of the problem space (Lesgold et al. 1988). In this way, the trainee was coached through the task, in a similar way to what would be offered by a mentor or private tutor. (Private tutoring has learning advantages over classroom instruction. It is as much as four times more effective and 98% of students perform better with private tutors; Bloom 1984.)

Sherlock was able to personalize instruction by creating an early version of a student model. The premise was that information must be known about students in order to tailor the instruction to their needs. With this in mind, Sherlock kept track of two levels of student model: a competence model and a performance model. The *competence model* recorded all of the trainee's interactions with Sherlock on previous instructional goals and could be used to determine which problem the learner would be presented with. The *performance* model determined how the problem would be delivered to that particular trainee.

It is likely that ITSs will eventually become a component of training for all aviation professionals. It is recommended that ITSs be regarded as a complement and extension, rather than a replacement, of existing classroom, simulator and real-world teaching practices. Although the technology is still evolving, this approach has significant potential to evolve teaching practice in our industry.

Conclusion

Our society is in the midst of a technological revolution where advancements and new applications appear on a daily basis. As technology has such a direct impact on our daily lives, it will probably not be long before new devices and methodologies make their way into our training centers.

Big data is likely to result in training that targets specific organizational issues; wearable technology will inform instructors about the state of their students; augmented reality will make its way into our workplaces, classrooms and simulator sessions; and ITSs will reinvent our understanding of e-learning as something entirely customized to our needs and our unique learning styles.

Yet it is crucial to remember that new technology does not necessarily result in more effective training. For example, in the earliest days of computer-based training, many of the organizations who were eager to be among the first to deploy e-learning produced very poor, ineffective courseware, because little was known at that time about how to make this type of training effective (Kearns 2010). To learn from the mistakes of the past, and to avoid repeating them, it will be crucial to adopt a cautious approach to the incorporation of future technology, ensuring (before it is deployed) that the new training method effectively improves learning.

Chapter 11

The Future of
Competency-Based Education in Aviation

Questions Considered in this Chapter:

- What will be some of the key issues in competency-based education in the future?
- What could be some of the solutions to these issues?
- How does the aviation industry grapple with the old and emerging areas of safety science?
- How can academics assist in the development of competency-based education?

Introduction

Competency-based education (CBE) is, in many ways, a logical answer to problems with training in the aviation industry. The industry is global in scale, which means that training for it needs to contend with the fragmenting pressures created by crossing multiple jurisdictions, languages, cultures and organizations. The identification and documentation of professional competence seems to be an obvious way of dealing with the diversity of a global industry. Written competencies appear to be an ideal vehicle for capturing, preserving and communicating what learners require entering their chosen professions or to maintain and enhance their skills. The competency-based approach is also appealing to regulators working in a critical, high-risk industry. Competencies specify professional knowledge, skill and attitude which give regulators a way to monitor, manage and control training to achieve benchmarks acceptable to stakeholders. Credentialing within such a system can also be consistent, efficient and effective. CBE is thus a model that seems to provide answers to several vexing concerns of the industry.

CBE also has the apparent advantage of being an innovative approach to education and training. Since achievement in CBE is measured by demonstration of outcomes, rather than time served in a training program (i.e., the traditional hours-based training approach in aviation), the model is flexible and potentially allows a more capable learner to graduate faster and a less capable learner to have more time to develop. Neither of these possibilities is a real option in a traditional time-served system. Also, as CBE is focused on learning outcomes, it does not matter what kinds of learning experiences lead to competent performance.

Thus, professional experience can lead to competence as readily as a formal training program. Overall, the CBE model appears to offer a flexible, learner-centered method and potential financial savings if training programs can be shortened. The benefits of the latter possibility are not lost on stakeholders with an interest in containing the very substantial costs of aviation initial and recurrent training.

However, as pointed out in Chapter 1, there are problems with CBE. "Competence" tends to be understood in a narrow way as a predetermined level of knowledge, skill and attitude for a certain job task. This conceptualization of competence as a unitary state does not align with the notion that skilled activity can be a continuum between novice and expert levels of performance. In CBE, competence is also envisaged as an individual attribute. However, contemporary learning theory and understandings of work emphasize the social nature of competence. CBE also considers competence to be something that can be reduced to textual form. Yet research into human expertise suggests that only a certain amount of performance can be translated into text (typically what can be visually observed or verbally described by the expert). The bulk of what constitutes expert performance is tacit (unspoken) and thus difficult to transcribe in written language. This is a formidable challenge for CBE, as there is a tendency for competencies to be reductionist—meaning they are an oversimplified version of true professional competence. This is more problematic when the "hermeneutic" work of the training designer, trainer and assessor is factored in. Hermeneutics refers to the art of interpreting and understanding written language and other artifacts. Training professionals are responsible for interpreting competencies, but diverse interpretations are often the result of the hermeneutic process. Thus, the formulation *and* interpretation of competency texts confront users of CBE with very significant problems.

Given the need to promote consistent training in diverse settings, and the perception that traditional training approaches leave room for improvement, how can the aviation industry envisage training in the future? How can the purported benefits of CBE be preserved while overcoming the identified problems with this approach? For the authors of this book, this is *the* fundamental question confronting an industry that appears to have committed to CBE. In the rest of this chapter we wrestle with this question. We do not have a definitive answer, but we suggest ways the problems of CBE might be reduced.

Competence

From the perspective of CBE, competencies are often written in a way that simplifies the complexity and richness of competent performance. We have defined competence as the ability to fully participate in a complex social practice. This definition deliberately avoids narrowing and highlights the complexity and sociality of competence. In this section we identify the especially problematic aspects of a narrow view of competence for the aviation industry. First, the "unitary" perspective on competence assumed by CBE is problematic for an industry that must reckon with the development of expertise through a continuum from novice to expert. Second, the individualistic perspective on competence is an issue in an industry in which competence is often dependent upon teams and groups, and where it does not always make sense to view competence as an individual attribute.

The Need to View Competence in Terms of a Continuum (Novice to Expert)

In Chapter 1 we outlined some of the perceived strengths and weaknesses of competency-based education. Throughout the book we have also presented many instances where professionals in the industry were having difficulty applying competency texts to professionals with differing levels of expertise, such as between a new pilot compared with an experienced airline captain. This issue is not new.

As far back as 1911, workers in the aircraft engineering profession identified that there were vast differences between the performance/quality of the work being carried out; some engineers were extremely effective in their tasks, while others were plain dangerous. At about the same time, Frederick Taylor (1904/1964) developed the classic time-and-motion study, which was used to identify what techniques and processes good aircraft maintenance engineers used to accomplish their work. By identifying good practice, it was possible to "set up systematic training and development activities that yielded improvements in workers' competence and consequently, increased effectiveness in organizations" (Sandberg 2000, 10). Even recent studies show that superior performances at work are "usually a result of specific sets of competencies combined in a particular way" (10).

The effects of Taylor's work are still embedded in practice today. However, just following process bypasses the issue of competence and assumes that the development toward expertise is linear—that to become competent, learners just need to be taught the correct steps. This thinking has led competency-based approaches to education and training to make simplistic assumptions about the nature of competence.

To make the issue more poignant, we are therefore differentiating levels of competence; in fact, many may differentiate between competence and expert performance, with a not-so-linear process between them. We must acknowledge that the levels of expertise between a newly hired crewmember and an airline pilot flight examiner are different. One could be viewed as competent while the other would be deemed an expert.

For years, the concept of expertise has been the focus of research and scholarly work in the endeavor to better understand education. For example, in some professions, competence is viewed as a range from competent novice to expert. Here "competence inventories" are developed to differentiate levels of performance, where competence must be viewed as more than just mere technical proficiency. For instance, staff would be categorized as follows:

- those who are technically able to perform a task but have very limited practical experience of actually doing so (e.g., the company could use them in an emergency or, if necessary, for a one-off activity);

- those who have successfully performed the task on a small number of occasions (e.g., the company could use them if the intention was to develop their expertise further, in a support role or if time is not necessarily a key criterion);
- those who have performed the task many times and under a variety of conditions (i.e., experienced worker standard; they are completely reliable);
- those who have substantial experience but are also able to support the learning of others (i.e., they can perform a coaching or mentoring role); and
- those who are world class, that is, they are able to think through and, if necessary, bring about changes in the ways that tasks are tackled (Brown 2010, 331).

As Morgan (1988) argued, the concept of competence encourages us to think not only about knowledge itself, but also about the knowledge that is required for competent work performance. It must be acknowledged that moving along the continuum does not occur for some. Not all those who start an air traffic control program pass and not all co-pilots pass command training to become a captain.

The industry should avoid commitment to a unitary view of competence, but this does not rule out such an approach either. A unitary view of competence is promoted in most models of competency-based education. It emerges in competency texts that mostly offer a single picture of performance, in training strategies that aim to develop the ability to undertake the one performance described in the competency and in assessment results of "competent" or "not competent." We argue, however, that the aviation industry needs to work with more nuanced conceptualizations of competence and that expertise models (such as Dreyfus and Dreyfus's (1986) model) offer a more appropriate view. The industry should be looking at "full participation in a complex social practice" in terms of continua and be very clear about moving people towards expertise.

The Need to View Competence as Both Individual and Social in Nature

One of the greatest challenges the industry will have to face in the future is to determine whether competence is to be viewed as an individual or team skill. For decades the industry has promoted training in soft skills, crew resource management and, more recently, non-technical skills (see Flin and Maran 2004; Flin et al. 2009). Yet pilots, engineers and air traffic controllers in every country of the world hold licenses as individuals. We have already seen a move toward assessments of non-technical skills, though when first introduced, a technical performance issue must also be attributed. Further investigations and research need to clearly come to grips with performance in its totality, rather than separated by specific constructs. CBE has the potential to further isolate how the aviation profession regards competence.

Competencies

The weakest link in any competency-based education system is probably the competency texts themselves. In this book we highlighted the problem of reductionism—an unavoidable issue when any attempt is made to translate competence or expertise into textual form. There must be loss, distortion and possibly misrepresentation as well when a cognitively and affectively complex, embodied form of knowledge is committed to a written statement. As the performance becomes increasingly complex, this distortion increases.

The aviation industry, with its imbedded complexity, requires that competencies written at an international level are as consistent as possible across multiple locations, organizations, jurisdictions and languages. This is necessary for effective regulation as well as education and training. Textual representation presents an obvious medium for promoting consistency and it is understandable that the industry is interested in documenting as many competencies as possible for the sake of consistent practice on the job and in education and training. We therefore have a significant challenge: how to provide a basis for consistent practice in a highly diverse environment while avoiding oversimplification within written statements.

Two approaches may assist in this challenge. First, assuming that some degree of textual representation is necessary in aviation, we propose a strategy of minimizing reliance on texts. The most obvious way to do this is to boost the amount of social sharing of practices. The most complex occupations can be effectively passed on in a social context (Lave and Wenger 1991). The social medium is not well understood, but it certainly is an effective way to teach and develop. In a geographically distributed industry, face-to-face social sharing may not be possible, yet other ways to tap into this approach must be considered. Social media and face-to-face meetings of representatives of different sites of practice may allow for social sharing for learning. For instance, in the Tasman region between New Zealand and Australia, there has been an increased level of sharing of information between airlines—even those that are commercial competitors—associated with concepts of competence, competencies and competency-based education. Use of videos of differing levels of performance in the flight deck are shared and discussed across the airlines to obtain a clear understanding of what performance looks like. This social interaction occurs at the top level, such as training managers and chief pilots, but is also filtered down to the newest pilots within the organization.

The second approach is to employ indigenous formal models of practice as a basis for writing competencies when these texts are used to promote consistency. These are sometimes called "folk models" to designate the fact that the model is generated from within an occupation. If reduction is to occur, because we need some level of textual representation of expertise, it may be better for it to take place using formalizations that have been invented by and for the industry. A couple of approaches used in aviation seem to be able to work not only with linear competencies but in ways that assist in understanding (representing) non-linear competencies. For example, within the European air traffic control system a model was developed that helped trainers to conceptualize—even at the most abstract level—how performance may be viewed. Here performance is not viewed as linear but rather as a group of interrelated competencies. In the model developed by Oprins and Schuver (2003; Oprins et al. 2006 is the English-language version), the outcomes—such as conflict resolution, delay time separation and communication—can be accurately assessed and the assessment can be semi-automated. However, competencies of an air traffic controller also consist of cognitive processes, plus emotional and social constructs. Therefore, controllers are assessed not only on their technical performance in completing the job, but also on the non-technical skills they have been taught (Oprins et al. 2006). As mentioned in Chapter 4, an experienced training manager outlined that "the instructors who are really good at this quite often would be able to work out where the deficiency was. ... [But] it only worked when the instructors really knew, really understood the model."[1] That is, instructors and assessors need to be able to recognize that a technical issue (e.g., a separation issue) may in fact be due to problems of coordination with another controller (Figure 11.1).

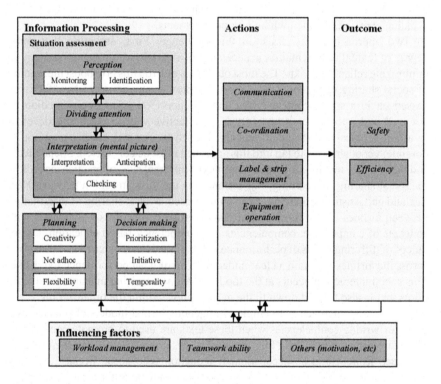

Figure 11.1   The ATC Performance Model
Source: Oprins et al. 2006, 305.

---

  1   ATC1: Air Traffic Control Trainer, Southeast Asia.

In pilot training, similar models have been developed to assist in understanding the integration of technical and non-technical skills. As outlined, non-technical skills have become an essential component of pilot training and assessment. However, once again, if we treat the development of competencies as a linear process, this does not allow for trainers and trainees to be able to discuss the complex interaction between technical and non-technical skills, especially in relation to complex aircraft flight decks. Mavin (2010) argued that, when assessing complex performance of an airline captain, experienced flight examiners do not separate performance according to a technical–non-technical divide. Rather, they see performance in terms of essential skills and enabling skills. Here the argument was based on the concept of compensatory and non-compensatory skills (e.g., Brannick and Brannick 1989). That is, senior pilots over a period of time develop an understanding of performance, where certain skillsets—such as situational awareness (cognition), aircraft flown within parameters (technical skill) and decision-making (non-technical skill)—were of the utmost importance (non-compensatory). However, deficiencies in areas such as technical knowledge (technical skill), management (non-technical skill) and communication (non-technical skill) influenced how well a pilot might perform (compensatory). The original model for assessing pilots' performance (MAPP) was a first attempt at a conceptual framework to provide a relationship between areas of competence. In the original MAPP, Mavin (2010) argued that the concept of teamwork was not a competency in itself but rather the integration of individual pilots' MAPP competencies (Figure 11.2).

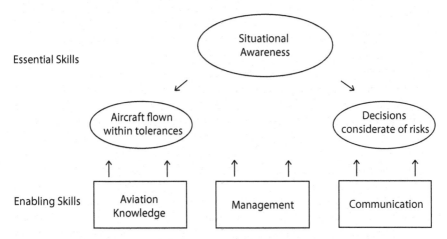

Figure 11.2   Original Model for Assessing Pilots' Performance (MAPP)
Source: Mavin and Dall'Alba 2010, 98.

Competency-Based Education in Aviation

Figure 11.3   Model for Assessing Pilots' Performance (MAPP)
Source: Mavin and Roth 2014, 221.

Later, Mavin and Roth (2014) made a case that an important component of complex work competencies involved understanding language use—a methodology they referred to as discursive psychology—to discuss performance, as much as the measurement of performance itself. They had modified the MAPP in a way that depicted each competency as a repertoire. Figure 11.3 shows how the MAPP is used to facilitate an understanding of complex performance. In the figure, (a) depicts the MAPP as a combination of six repertoires: situational awareness, tolerances (aircraft maintained within the specific tolerances required by the airline), decision-making, aviation knowledge, management and communication. Figure 11.3 (b) may be exemplified by a flight attendant relaying to the captain that a passenger has become ill (communication). The captain becomes aware of the situation (situational awareness) and makes a decision to divert the aircraft to an alternate aerodrome. In (c), the captain will then identify what tasks are required, the priority of each task and finally the delegation of each task (management). Here it is possible to see that a debriefing of a pilot simulator session would include an understanding of the relationship between each of the repertoires.

Mavin and Roth (2014) advanced the understanding of performance from a practitioner's perspective in two ways. First, the fundamental argument about understanding the language used to discuss performance was that it "sidesteps" many of the current arguments associated with areas such as measuring inter-rater reliability, how one goes about measuring cognition, measuring repertoires such as situational awareness or even ideas associated with psychometrics:

> By asking questions related to the usefulness of particular discourses in dealing with the everyday world, even if raters differ in the way in which they categorize performance—for example, whether it was knowledge or situational awareness that was "weak" in a particular situation—the interpretive repertoires are never at stake. These are taken as the ground that never comes to be questioned in pilot-assessment-related discourse. That is, the discursive approach is concerned with the ways in which everyday people, such as flight examiners, constitute and talk about the worlds of their experience. It is also concerned with the ways in

which categorical terms (e.g., "good," "poor," "minimum standard") are used to assign qualities to the performance dimensions. (Mavin and Roth 2014, 223)

The second argument they made was that of teamwork being a combination of repertoires within individuals:

> To use an analogy, then, MAPP constitutes something like a collection of primary colors from which other colors can be produced. Teamwork therefore defines the required color, and consequently, each specific situation—say normal versus emergency—might require a differing blend of primary colors to satisfy the situation. For example, a captain who faces a scenario where his lack of knowledge (knowledge) of a system might be impeding his ability to diagnose the situation (situational awareness and decision making) might ask the first officer for his or her take on the situation (communication) or delegate (management) the flying of the aircraft (aircraft maintained within tolerances) so a manual (knowledge) can be consulted. This is good teamwork, yes, although described using a selection of the six repertoires. The repertoires described earlier constitute a set of fundamental discursive means to talk about every aspect of performance in the cockpit. (Mavin and Roth 2014, 224)

Industry insiders close to the intricacies of occupational practices, for example, have developed models such as those put forth by Oprins and colleagues and by Mavin and colleagues. Such models are likely to be a better basis for constructing textual representations of practice at more complex levels than generic behavioral models that come from outside the industry. As a model from outside the industry, the elements/performance criteria conceptualization used to construct competencies has no affinity with indigenous formalizations (i.e., produced by and for the industry). However, using indigenous models at least allows more sophisticated representation that is closer to actual occupational practices. Thus the Oprins model articulates air traffic control and the Mavin model articulates pilot work in a way that allows features unique to these occupations to be represented. It must be stressed that the models discussed are only first attempts. Using these frameworks as a starting point, or even using the ideas discussed and starting all over again, may assist the industry to avoid oversimplification and develop competencies that better embody the complex attributes of expert performance.

If texts used for training and regulation are based on these models, reliance on such texts is not as problematic. In addition, if the continued evolution of these models—and competing models—is something training designers, trainers and assessors are invited to participate in, the texts themselves become living documents subject to interrogation and change. The challenge of formalization then becomes a shared concern rather than a process that produces texts that are regarded as alien objects by educators and trainers.

Even though the use of conceptual frameworks may assist in developing a better understanding of complex performance, they are certainly not without criticism. Models, or "folk models" as they are often called, are generally at a conceptual level only. However, the models themselves are generally made up of constructs that are also conceptual. "The problem is rather that the value of the constructs hinges on their common-sense appeal rather than their substance" (Dekker and Hollnagel 2004, 80). As Dekker and Hollnagel (2004) explain quite clearly, there are limitations that must be taken into account with regard to conceptual models:

> The most evident characteristic of folk models is that they define their central constructs, the *explanandum*, by substitution rather than decomposition or reduction. So instead of explaining the central construct by statements that refer to more fundamental and presumably better known *explananda*, the explanation is made by referring to another phenomenon or construct that itself is in equal need of explanation. A good example is complacency, as it is used in relation to the problems observed on automated flight decks. Most textbooks on aviation human factors talk about complacency and even endow it with causal power, but very few define it. (81)

One way to allow for the strengths and inherent weakness of models is to use a mixture of both. For example, one airline using MAPP, only uses traditional competencies in textual form during early training. This allows for detailed criteria needed during early instruction, generally centered on technical skills development. As a new pilot progresses along a training program, skill sets become increasing complex, integrated and social in nature (two pilot crew). Accordingly, the number of competencies required to represent such performance had become detailed, onerous and impractical to use. Here the MAPP is brought into play with competencies written around the dimension. In fact, in the latter part of training the competencies have moved from 56 pages to one. This approach has allowed for a more seamless and practical training system.

In summary, how an individual, an organization or even the regulator understands competence must take into account not only the linear competencies associated with performance but also an understanding of performance at a far more complex level.

Competency-Based Education

We have defined competency-based education as "instructional design, training, and assessment that systematically references competency texts." We have drawn attention to a significant issue in this conceptualization of education—that reference to competency texts is in fact a complex hermeneutic process. Expertise in an occupation is no guarantee of the ability to decode competency texts. Dreyfus and Dreyfus's (1986) model of expertise describes the highest reaches of expert performance as entailing a lack of reference to explicit rules. Experts do not employ translations or codifications of their work to do the work. In contrast, novices do refer to formal rules when they engage in the activities of the work. Hodge (2014a) argued that this might account for the fact that the experts in his research did not relate well to competency texts. That is, the texts represent their work in a way that appears foreign to experts. If experts are employed to design and conduct training and assessment, they also need to learn how to interpret competency texts. Being an expert provides *no* privileged access to competencies. The hermeneutic skills of interpretation are required, in addition to occupational expertise, to work effectively with the competency-based approach.

To confront the hermeneutic challenge, a number of measures may be effective. Two have already been proposed. Reduction of reliance on texts by increasing opportunities for social sharing of practices may serve to build hermeneutic skills. Sharing and defending interpretations appears to be an effective way to develop hermeneutic expertise (Hodge 2014b) and social sharing can also present opportunities to share interpretations of whatever texts are used. If contributing to the continued development of formal models for representing expertise is part of the work of educators and trainers, these practitioners will have greater investment in the structure of the texts they work with and they will increase their hermeneutic expertise.

Some Future Implications for CBE

Dealing with the Current Theories of Safety

In addition to viewing competency-based education as competence, competencies and competency-based education, there are issues that will need to be resolved in the future in regard to safety. As we have described, some performance requirements within aviation are complex, because in addition to individual skills, professionals are required to possess the teamwork skills to work in a group. Additionally, over the last decades, performance in industries such as nuclear power, medicine and aviation have been viewed as occurring in a complex sociotechnical system. That is, the aviation system is made up of many human and machine agents that work collaboratively. Even the smallest change in one part of the system can create an unforeseen accident somewhere else.

Over the years, the aviation industry has been exposed to three main areas of safety science: normal accident theory, high-reliability organization (HRO) and, more recently, resilience engineering. Normal accident theory was developed in the 1980s and explains how unexpected failures occur within "tightly coupled" systems where humans and machines are highly dependent on one another. Accidents within such complex systems are both unforeseen and unfortunately unavoidable (see Perrow 1999).

On the other hand, HRO is characterized as follows:

a. all participating organizations in the network rely on each other to provide error-free contributions to the overall performance;
b. not all organizations participating in the network are necessarily high-reliability organizations used to dealing with high-risk environments;
c. the failure of one participant may thwart the reliability of the whole network's performance;
d. reliable combination of all organizational contributions to the overall performance of the network is supported via specific cooperation structures and practices that aim at integration, also of a cultural nature, at the network level. (Modified from Berthod et al. 2015, 26)

The last area of safety science is that of resilience engineering. Resilience engineering supporters have argued, quite convincingly, that current models do not allow for the skills that allow people to deal with unforeseen events, which are characteristics of HRO. Simply focusing on errors that people make, and trying to avoid them as HROs do, is problematic:

> The main question should not be why accidents occur and whether people committed 'errors', but what people do in order to prevent and avoid accidents. Therefore, our focus is more on the dynamic 'navigation' of the system in the face of changing challenges and not on momentary events, which are minimally emblematic of change. Risks and hazards, therefore, always exist in operations and maintenance activities; some are known, others are not. (Carayon et al. 2015, 555)

One thing that we must be careful of is that all of the approaches have received criticism from each other. Hopkins (2014) discussed:

> the problematic nature of some of the most widely referenced theories or theoretical perspective in our inter-disciplinary field, in particular, normal accident theory, the theory of high reliability organisations, and resilience engineering. Normal accident theory turns out to be a theory that fails to explain any real accident. HRO theory is about why HROs perform as well as they do, and yet it proves to be impossible to identify empirical examples of HROs (beyond those originally studied) for the purpose of either testing or refining the theory. Resilience engineering purports to be something new, yet on examination it is hard to see where it goes beyond HRO theory. (13)

While normal accident theory and HROs have been around for some time, and have received their fair share of critiques, resilience engineering is the new kid on the block. The most obvious criticism of resilience engineering so far is that the high levels of safety we enjoy today in aviation are solely due to the *old ways* (like HRO). Critics generally claim that the conceptual approach of resilience engineering takes the safety field little further than the work already done in the late 1980s and 1990s in high-reliability theory (HRT). Also, it appears that while resilience engineering appears to be gaining increasing support, it has yet to be situated within the academic community. Many of the publications are either conference papers or books, and not peer-reviewed articles in academic journals:

> Even though the resilience engineering research agenda can hardly be called new (nor heretic), research concerning the topic still does not seem to aim for peer-reviewed journals as their main outlets. From our literature searches, it becomes evident that the field of resilience engineering has yet to position itself in the wider peer-reviewed scientific community. (Bergström et al. 2015)

We are, as educators, always interested in these philosophical debates. Not for reasons of who is correct and who is not, as that is not our area. We simply need to know how to derive our theories for competence, define competencies and teach them, thus enabling us to provide clear guidelines into what the industry requires us to deliver. The aim of this section is just to outline some of the theoretical perspectives in safety science that are currently out there in the field of aviation (and other industries). Thus it is important when we talk about competence-based education *itself* that we have an understanding of some of the theoretical ideas and perspectives currently going around.

A Final Note on How the Research Community Can Help

Only recently the argument was made that "academic institutions and airlines have always worked together to develop and conduct research studies. However, most often the expertise or areas of interest of the academics have driven these studies" (Mavin et al. 2015, 52). It is imperative that for research to become more effective we must bring together both industry and aviation researchers. Yet, are we our worst enemy? As one of the professionals we interviewed explained:

> We will need to combine an understating of competencies, human factors, SMS, CRM, etc. We need to keep this alive. The other piece I found, if you talk to certain pilots they'll say, "yeah, but that's academic." There needs to be a better link between applied research and the industry. There needs to be a better understanding between those two communities and I think there is still a lot of, maybe mistrust, but there's a lot of "you don't know what I do—I'll tell you how to teach me." There's a lack of recognition for the type of expertise needed to understand CBT and it's not coming from the people actually doing it. [For example, you need inspectors] who can look at big data. Build a picture, and understand what the trend is. For [CBE] there's a similar kind of need to draw out expertise from professionals and bring it into training. But, to do that, we need to be allowed to get close to them. Maybe it's happening. (PIL9)[2]

---

2   PIL9: Instructional Designer, North America.

We believe a way forward includes three important points. First, many of those within the academic community are very much aware that they do not have the experiences of those in the field (though we must be also careful with those who do, as they can end up with some type of god status). However, it is incumbent upon industry to move forward to assist many academics to gain insight into practice, thereby enabling the academics' own understanding of professional practice and current issues. That way the academic can align to current issues. As one of the authors of this text asked, "What is keeping professionals up at night? What is the training manager worried about? Not sure how we might investigate the problem, but it sure as hell is what we should be focusing on."

Second, academics should endeavor to understand that multiple methods might be used to investigate a particular problem. While we understand and appreciate the rigor associated with some of the quantitative methodologies available, some of the greatest insights into learning have occurred through small-scale qualitative methodologies. For example, work by John Dewey on developmental psychology over half a century ago has been extremely influential—even today—in our understanding about child development. Dewey's original study had three participants: his own three children. We have observed qualitative researchers being snubbed by other researchers at conferences, or being told by their own university colleagues that their research is based on "small sample size" or that their methods are "epistemological crap." We believe that both qualitative and quantitative research approaches are necessary for progress, particularly within competency-based education as it is a relatively new field of exploration for aviation researchers.

Finally, academics should do the best they can—and this must be assisted by the aviation industry—to try to explain research findings in ways that allow professionals within the industry to understand and implement findings. Our research work must bridge the theory–practice gap that is so prevalent in academia (Roth et al. 2014). It is hoped that academics can help bridge the theory–practice gap by translating research findings into language that is accessible to practitioners.

Conclusion

The aim of this chapter was to outline some of the issues that need to be considered when moving forward into the future with regard to competency-based education. While throughout the book we have outlined some of the strengths of competency-based education, there are also some challenges that must be met by the aviation industry as a whole. This includes not only the industry, but also the academic community where scholarly input can be of huge benefit.

Part IV

References

Adams, R. 2009. 27 into 7 equals 9625 – The new regulatory matrix. *Civil Aviation Training Magazine* 1: 12–15.

Adams, R. 2010. *A History of Simulation: Part 1.* September 2. Retrieved from Inside S&T Defence: http://halldale.com/insidesnt/history-simulation-part-i#. VO9KrctOxMs. Accessed February 25, 2015.

Azuma, R.T. 1997. A survey of augmented reality. *Presence* 6(4): 355–85.

Azuma, R., Y. Baillot, R. Behringer, S. Feiner, S. Julier and B. MacIntyre. 2001. Recent advances in augmented reality. *Computer Graphics* November/December: 34–47.

Baker, R.S., A.T. Corbett, K.R. Koedinger, S. Evenson, I. Roll, A.Z. Wagner, et al. (n.d.). *Adapting to When Students Game an Intelligent Tutoring System.* Retrieved from Columbia University: http://www.columbia.edu/~rsb2162/Baker175.pdf. Accessed September 16, 2014.

BBC. 2015. *Google Glass Sales Halted but Firm Says Kit is Not Dead.* January 15. Retrieved from BBC News Technology: http://www.bbc.com/news/technology-30831128. Accessed February 16, 2015.

Bergström, J., R. van Winsen and E. Henriqson. 2015. On the rationale of resilience in the domain of safety: A literature review. *Reliability Engineering & System Safety.* doi:10.1016/j.ress.2015.03.008.

Berthod, O., M. Grothe-Hammer and J. Sydow. 2015. Some characteristics of high-reliability networks. *Journal of Contingencies and Crisis Management* 23(1): 24–8.

Bloom, B.S. 1984. The 2 sigma problem: The search for methods of group instruction as effective as one-to-one. *Educational Researcher* 13(6): 4–16.

Bom, B.S. 1984. The 2 sigma problem: The search for methods of group instruction as effective as one-to-one. *Educational Research* 13: 4–16.

Brannick, M.T. and J.P. Brannick. 1989. Nonlinear and noncompensatory processes in performance evaluation. *Organizational Behavior and Human Decision Processes* 44(1): 97–122.

Brown, A. 2010. Assessment in the workplace of performance, developing expertise and competence. In *International Encyclopedia of Education*, eds P. Peterson, E. Baker and B. McGaw, 3: 330–36. Oxford, UK: Elsevier.

Butler, D. 2013. *When Google Got Flu Wrong.* February 13. Retrieved from http://www.nature.com/news/when-google-got-flu-wrong-1.12413. Accessed February 16, 2015.

Capuano, N., M. Marsella and S. Salerno. 2000. ABITS: An agent-based intelligent tutoring system for distance learning. *Proceedings of the International Workshop on Adaptive and Intelligent Web-based Educational Systems.* Montreal, Canada: International Conference on Intelligent Tutoring Systems.

Carayon, P., P. Hancock, N. Leveson, I. Noy, L. Sznelwar and G. van Hootegem. 2015. Advancing a sociotechnical systems approach to workplace safety—developing the conceptual framework. *Ergonomics* 58(4): 548–64.

Cheung, B., L. Hui, J. Zhang and S.M. Yiu. 2003. SmartTutor: An intelligent tutoring system in web-based adult education. *The Journal of Systems and Software* 68: 11–25.

Cisco. 2014. *Cisco Visual Networking Index: Forecast and Methodology, 2013–2018.* June 10. Retrieved from Cisco: http://www.cisco.com/c/en/us/solutions/collateral/service-provider/ip-ngn-ip-next-generation-network/white_paper_c11-481360.html. Accessed February 16, 2015.

Cisco. 2015. *Cisco Visual Networking Index: Global Mobile Data Traffic Forecast Update, 2014–2019.* February 3. Retrieved from Cisco: http://www.cisco.com/c/en/us/solutions/collateral/service-provider/visual-networking-index-vni/white_paper_c11-520862.pdf. Accessed February 16, 2015.

Clark, R.E. 2012. *Learning from Media: Arguments, Analysis and Evidence.* 2nd edn. Greenwich, CT: Information Age Publishing.

Crowther, B.T. 2012. (Un)reasonable expectation of digital privacy. *Brigham Young University Law Review* 1: 343–70.

Dekker, S. and E. Hollnagel. 2004. Human factors and folk models. *Cognition, Technology & Work* 6(2): 79–86.

de Winter, J.C., D. Dodou and M. Mulder. 2012. Training effectiveness of whole body flight simulator motion: A comprehensive meta-analysis. *The International Journal of Aviation Psychology* 22(2): 164–83.

D'Mello, S.K., S.D. Craig, B. Gholson and S. Franklin. 2005. Integrating affect sensors in an intelligent tutoring system. *Affective Interactions: The Computer in the Affective Loop Workshop at 2005 International Conference on Intelligent User Interfaces,* 7–13. New York: ACM Press.

Dreyfus, H.L. and S.E. Dreyfus. 1986. *Mind over Machine: The Power of Human Intuition and Expertise in the Era of the Computer.* New York: Free Press.

Duval, S. and H. Hashizume. 2005. Perception of wearable computers for everyday life by the general public: Impact of culture and gender on technology. *Embedded and Ubiquitous Computing* 3824: 826–35.

Flin, R. and N. Maran. 2004. Identifying and training non-technical skills for teams in acute medicine. *Quality and Safety in Health Care* 13(1): 80–84.

Flin, R., P. O'Connor and M. Crichton. 2009. *Safety at the Sharp End.* Aldershot, UK: Ashgate.

Frasson, C., T. Mengelle and E. Aimeur. 1997. Using pedagogical agents in a multi-strategic intelligent tutoring system. *Proceedings of the AI-ED Workshop on Pedagogical Agents,* 40–47.

Freedman, R. 2000. What is an intelligent tutoring system? *Intelligence* 11(3): 15–16.

Gegenfurtner, A., C. Quesada-Pallares and M. Knogler. 2014. Digital simulation-based training: A meta-analysis. *British Journal of Educational Technology*, 1097–114.

Harrop, P., J. Hayward, R. Das and G. Holland. 2015. *Wearable Technology 2015–2025: Technologies, Markets, Forecasts.* IDTechEx.

Hodge, S. 2014a. Expertise and the representation of knowledge in training packages. Paper presented at the AVETRA 17th Annual Conference, Surfers Paradise, QLD. http://avetra.org.au/wp-content/uploads/2014/05/Abstract-46. pdf. Accessed November 11, 2015.

Hodge, S. 2014b. *Interpreting Competencies in Australian Vocational Education and Training: Practices and Issues.* Adelaide, SA: NCVER.

Hoofnagle, C.J., J. King, S. Li and J. Turow. 2010. *How Different are Young Adults from Older Adults When it Comes to Information Privacy Attitudes and Policies?* April 14. Retrieved from SSRN: http://dx.doi.org.proxy1.lib.uwo. ca/10.2139/ssrn.1589864. Accessed January 23, 2015.

Hopkins, A. 2014. Issues in safety science. *Safety Science* 67: 6–14.

Huba, M.E. and J.E. Freed. 2000. *Learner-centered Assessment on College Campuses: Shifting the Focus from Teaching to Learning.* Needham Heights, MA: Allyn & Bacon.

Iachello, G. and J. Hong. 2007. End-user privacy in human-computer interaction. *Foundations and Trends in Human Computer Interaction* 1(1): 1–137.

International Civil Aviation Organization. 2003. *Manual of Criteria for the Qualification of Flight Simulators.* Montreal: ICAO.

Kearns, S. 2010. *e-Learning in Aviation.* Aldershot, UK: Ashgate.

Kirkham, R. and C. Greenhalgh. 2015. Social access vs. privacy in wearable computing: A case study of autism. *IEEE Pervasive Computing* January–March: 26–33.

Lave, J. and E. Wenger. 1991. *Situated Learning: Legitimate Peripheral Participation.* Cambridge, UK: Cambridge University Press.

Lesgold, A., S. Lajoie, M. Bunzo and G. Eggan. 1988. *Sherlock: A Coached Practice Environment for an Elecronics Troubleshooting Job.* Pittsburgh, PA: Learning Research and Development Center.

Li, L. and R.J. Hansman. 2013. *Anomaly Detection in Airline Routine Operations Using Flight Data Recorder Data.* Cambridge, Massachusetts: MIT International Center for Air Transportation (ICAT).

Licklider, J.C. 1960. Man-Computer Symbiosis. *IRE Transactions on Human Factors in Electronics, HFE-1* March: 4–11.

Long, C. 2014. Big data – the Ethiad Airways approach. *Civil Aviation Training* 5: 8–11.

Mackay, W.E., A.-L. Fayard, L. Frobert and L. Medini. 1998. Reinventing the familiar: Exploring an augmented reality design space for air traffic control.

Proceedings of the SIGCHI Conference on Human Factors in Computing Systems, 558–65. Los Angeles: ACM Press/Addison-Wesley Publishing Co.

Margaryan, A., A. Littlejohn and G. Vojt. 2011. Are digital natives a myth or a reality? University students' use of digital technologies. *Computers & Education* 56: 429–40.

Marsh, A.K. 2011. *ABCs of Simulators: A Tangled Web We Weave.* May 1. Retrieved from AOPA: http://www.aopa.org/News-and-Video/All-News/2011/May/1/ ABCs-of-Simulators. Accessed February 25, 2015.

Mavin, T.J. 2010. *Assessing Pilots' Performance for Promotion to Airline Captain.* Professional Doctorate: The University of Queensland.

Mavin, T.J. and G. Dall'Alba. 2010. A model for integrating technical skills and NTS in assessing pilots' performance. In *Proceedings of the 9th International Symposium of the Australian Aviation Psychology Association*, 9–12. Sydney, Australia.

Mavin, T.J. and W.M. Roth. 2014. A holistic view of cockpit performance: An analysis of the assessment discourse of flight examiners. *International Journal of Aviation Psychology* 24(3): 210–27.

Mavin, T.J., W.-M. Roth, K. Soo and I. Munro. 2015. Toward evidence-based decision making in aviation. *Aviation Psychology and Applied Human Factors* 5: 52–61.

MongoDB. 2014. *Big Data: Examples and Guidelines for the Enterprise Decision Maker.* August 1. Retrieved from MongoDB White Paper: http://info.mongodb. com/rs/mongodb/images/10gen_Big_Data_White_Paper.pdf?_ga=1.2338853 7.2120011512.1422459664. Accessed January 23, 2015.

Morgan, G. 1988. *Riding the Waves of Change: Developing Managerial Competencies for a Turbulent World.* San Francisco, CA: Jossey-Bass.

Moroney, W.F. and M.G. Lilienthal. 2009. Human factors in simulation and training: An overview. In *Human Factors in Simulation and Training*, eds D.A. Vincenzi, J.A. Wise, M. Mouloua and P.A. Hancock, 4–38. Boca Raton, FL: CRC Press.

Nielsen. 2014. *Tech-Styles: Are Consumers Really Interested in Wearing Tech on Their Sleeves.* March 20. Retrieved from Media and Entertainment: http:// www.nielsen.com/us/en/insights/news/2014/tech-styles-are-consumers-really-interested-in-wearing-tech-on-their-sleeves.html. Accessed February 18, 2015.

Norman, G., K. Dore and L. Grierson. 2012. The minimal relationship between simulation fidelity and transfer of learning. *Medical Education* 46(7): 636–47.

Ohm, P. 2012. The underwhelming benefits of big data. *University of Pennsylvania Law Review* 161: 339–46.

Oprins, E. and M. Schuver. 2003. Competentiegericht opleiden en beoordelen bij LVNL [Competence-based training and assessment at LVNL]. *HUFAG Nieuwsbrief* 6: 2–4.

Oprins, E., E. Burggraaff and H. van Weerdenburg. 2006. Design of a competence-based assessment system for air traffic control training. *International Journal of Aviation Psychology* 16(3): 297–320.

Palen, L., M. Salzman and E. Youngs. 2001. Discovery and integration of mobile communications in everyday life. *Personal and Ubiquitous Computing* 5(2): 109–22.

Park, S. and S. Jayaraman. 2003. Enhancing qualtiy of life through wearable technology: The role of a personalized wearable intelligent information infrastructure in addressing the challenges of healthcare. *IEEE Engineering in Medicine and Biology Magazine* May/June: 41–8.

Pentecost, M.J. 2015. Big data. *Journal of the American College of Radiology* 12(2): 129.

Perng, B., S. Fisher, S. Hollar and K.S. Pister. 1999. Acceleration sensing glove (ASG). *Proceedings of the Third International Symposium on Wearable Computers (ISWC)*, 178–80. San Francisco: IEEE.

Perrow, C. 1999. *Normal Accidents.* Princeton, NJ: Princeton University Press.

Picard, R. and J. Scheirer. 2001. The galvactivator: A globe that senses and communicates skin conductivity. *Proceedings on the 9th International Conference on Human-Computer Interaction*, 1538–42. New Orleans, LA.

Prensky, M. 2001. Digital natives, digital immigrants. *On the Horizon* 9(5): 1–6.

Rogers, E., R.R. Murphy and C. Thompson. 1997. Outbreak agent: Intelligent wearable technology for hazardous environments. *IEEE International Conference on Systems, Man, and Cybernetics*, 3198–203. Orlando, FL: IEEE Press.

Roscoe, S.N. 1991. Simulator qualification: Just as phony as it can be. *International Journal of Aviation Psychology* 1(4): 335–9.

Roth, W.M., T.J. Mavin and S. Dekker. 2014. The theory–practice gap: Epistemology, identity, and education. *Education + Training* 56(6): 521–36.

Roth, W.M., T.J. Mavin and I. Munro. 2014. How a cockpit forgets speeds (and speed-related events): Toward a kinetic description of joint cognitive systems. *Cognition, Technology & Work.* doi:10.1007/s10111-014-0292-0.

Rowther, B.T. 2012. (Un)Reasonable expectation of digital privacy. *Brigham Young University Law Review* 1: 343–70.

Royal Aeronautical Society Flight Simulation Group. 2009. *The Impact of Flight Simulation in Aerospace.* March. Retrieved from Royal Aeronautical Society: http://aerosociety.com/Assets/Docs/Publications/DiscussionPapers/The_impact_of_flight_simulation_in_aerospace.pdf. Accessed February 25, 2015.

Salas, E. and C.S. Burke. 2002. Simulation for training is effective when … . *Quality and Safety in Health Care*, 119–20.

Sana, F., T. Weston and N.J. Cepeda. 2013. Laptop multitasking hinders classroom learning for both users and nearby peers. *Computers & Education* 62: 24–31.

Sandberg, J. 2000. Understanding human competence at work: An interpretative approach. *Academy of Management Journal* 43(1): 9–25.

Setz, C., J.S. Arnrich, R. La Marca and G.E. Troster. 2010. Discriminating stress from cognitive load using a wearable EDA device. *IEEE Transactions on Information Technology in Biomedicine* 14(2): 410–17.

Sims, D. 1994. New realities in aircraft design and manufacture. *IEEE Computer Graphics and Applications* 14(2): 91.

Sitzmann, T. 2011. A meta-analytic examination of the instructional effectiveness of computer-based simulation games. *Personnel Psychology* 64: 489–528.

Smith, E.E. 2012. The digital native debate in higher education: A comparative analysis of recent literature. *Canadian Journal of Learning and Technology* 38(3): 1–18.

Strachan, I. 2014. World civil full flight simulator census. *Civil Aviation Training* 4: 60–61.

Taylor, F. 1904/1964. *The Principles of Scientific Management*. New York: Harper.

Tennyson, R.D. and R.L. Jorczak. 2008. A conceptual framework for the empirical study of instructional games. In *Computer Games and Team and Individual Learning*, eds H.F. O'Neil and R.S. Perez, 39–54. Oxford, UK: Elsevier.

The Economist. 2014. *Big Data: Turning the Tables*. September 3. Retrieved from The Economist, Business and Finance: http://www.economist.com/news/business-and-finance/21613499-returning-favour. Accessed January 23, 2015.

Ulrich, W.F. 2010. A history of simulation: Part 2 – early days. http://halldale.com/insidesnt/history-simulation-part-ii-early-days#.VkNFQrerTIU. Accessed November 11, 2015.

Vallurupalli, S., H. Paydak, S.K. Agarwal, M. Agrawal and C. Assad-Kottner. 2013. Wearable technology to improve education and patient outcomes in a cardiology fellowship program – a feasibility study. *Health Technology* 3: 267–70.

Van Laerhoven, K. and O. Cakmakci. 2000. What shall we teach our pants? In *The Fourth International Symposium on Wearable Computers*, 77–83. Atlanta: IEEE.

Warren, S.D. and L.D. Brandeis. 1890. The right to privacy. *Harvard Law Review* 4(5): 193–220.

Westin, A.F. 1967. *Privacy and Freedom*. New York: Atheneum.

Wickens, C.D., S. Hutchins, T. Carolan and J. Cumming. 2013. Effectiveness of part-task training and increasing-difficulty training strategies: A meta-analysis approach. *Human Factors* 55(2): 461–70.

Wright, R. and L. Keith. 2014. Wearable technology: If the tech fits, wear it. *Journal of Electronic Resources in Medical Libraries* 11(4): 204–16.

Wu, T., C. Dameff and J. Tully. 2014. Integrating Google Glass into simulation-based training: Experiences and future directions. *Journal of Biomedical Graphics and Computing* 4(2): 49.

Index

Page numbers in **bold** refer to figures and tables.

Printed in the United States
by Baker & Taylor Publisher Services